生物多样性与可持续发展系列丛书

黄山风景名胜区生物多样性保护行动计划（2018—2030年）

张丽荣 孟 锐 金世超 杨新虎 等著

中国环境出版集团·北京

图书在版编目（CIP）数据

黄山风景名胜区生物多样性保护行动计划（2018—2030 年）/
张丽荣等著. —北京：中国环境出版集团，2021.12
（生物多样性与可持续发展系列丛书）
ISBN 978-7-5111-4272-6

Ⅰ. ①黄⋯ Ⅱ. ①张⋯ Ⅲ. ①黄山—生物多样性—生物资源
保护—方针政策　Ⅳ. ①X176

中国版本图书馆 CIP 数据核字（2020）第 294922 号

出 版 人	武德凯
策划编辑	王素娟
责任编辑	王　菲
责任校对	任　丽
封面设计	彭　杉

出版发行　中国环境出版集团
　　　　　（100062　北京市东城区广渠门内大街 16 号）
　　　　　网　　址：http://www.cesp.com.cn
　　　　　电子邮箱：bjgl@cesp.com.cn
　　　　　联系电话：010-67112765（编辑管理部）
　　　　　　　　　　010-67162011（第四分社）
　　　　　发行热线：010-67125803，010-67113405（传真）

印　　刷	北京建宏印刷有限公司
经　　销	各地新华书店
版　　次	2021 年 12 月第 1 版
印　　次	2021 年 12 月第 1 次印刷
开　　本	787×1092　1/16
印　　张	10.25
字　　数	260 千字
定　　价	76.00 元

【版权所有。未经许可，请勿翻印、转载，违者必究。】
如有缺页、破损、倒装等印装质量问题，请寄回本集团更换

中国环境出版集团郑重承诺：
中国环境出版集团合作的印刷单位、材料单位均具有中国环境标志产品认证；
中国环境出版集团所有图书"禁塑"。

参与人员名单

领导组成员：

黄山风景名胜区管委会　吴文达　宋生钰　杨新虎

生态环境部环境规划院　王金南　陆　军　何　军　王夏晖

技术组成员：

黄山风景名胜区管委会　钱阳平　胡降临　叶要清　潘少杰　王　洋
　　　　　　　　　　　余　悦　包　强　位纯西　钱　为

生态环境部环境规划院　张丽荣　孟　锐　金世超　潘　哲　李若溪
　　　　　　　　　　　刘　洋　公滨南　周云峰

编写人员名单

总体统筹：张丽荣　钱阳平

规划统稿：张丽荣　孟　锐

专题研究：专题1　黄山风景名胜区生物多样性概况：
　　　　　　　　金世超　刘　洋

　　　　　专题2　黄山风景名胜区文化与旅游发展：
　　　　　　　　李若溪　孟　锐

　　　　　专题3　黄山风景名胜区生物多样性价值评估：
　　　　　　　　张丽荣　孟　锐　金世超

　　　　　专题4　黄山风景名胜区旅游环境容量评价：
　　　　　　　　孟　锐　李若溪

　　　　　专题5　黄山风景名胜区生物多样性保护空间分区：
　　　　　　　　潘　哲　刘　洋

制　　图：潘　哲　公滨南　周云峰

前　言

生物多样性是人类赖以生存和发展的必要条件，是维护国家生态安全的物质基础。自 1992 年达成并签订《生物多样性公约》以来，在各方共同努力下，世界生物多样性保护取得了良好进展。中国是世界上生物多样性最丰富的国家之一，中国政府高度重视生物多样性保护，将其作为生态文明建设的重要内容。近年来，在生态文明思想指引下，中国政府不懈努力，生物多样性保护主流化进程明显提速，社会参与度和公众意识不断提高，为全球生物多样性保护做出了积极贡献。但同时，我们也应该认识到，作为生物多样性受威胁最严重的国家之一，我国生物多样性加速下降的总体趋势尚未得到有效遏制。

黄山风景名胜区（正文中简称景区）位于安徽省黄山市，1982 年被国务院列为首批国家级重点风景名胜区，同时也是全国文明风景旅游区和国家 5A 级旅游景区。景区 1990 年被列入世界文化与自然遗产名录，2004 年成为世界地质公园，2018 年加入世界生物圈保护区网络。景区地处中亚热带北缘，地理位置独特，群山起伏，地貌多样，海拔高差 1 600 多米，是第四纪冰期动植物的避难场，生物多样性丰富，有 2 385 种高等植物和 417 种脊椎动物，占全国 0.044%的陆地面积上分布着全国 6.92%的植物物种和 9.55%的动物物种。近年来，景区通过创新景点"轮休"制度、加强古树名木保护、严格开展植物检疫、强化防火措施、积极保护水资源等重要措施，生物多样性保护能力显著提升。

为贯彻落实国家、省市的相关要求和部署，进一步加强景区生物多样性保护与管理工作，有效应对生物多样性保护面临的新问题、新挑战，受景区管委会园林局委托，生态环境部环境规划院组织技术团队编制了《黄山风景名胜区生物多样性保护行动计划（2018—2030 年）》（以下简称行动计划）。2019 年 5 月，黄山市人民政府批复同意实施行动计划，力争将景区建设成为全国山岳型景区生物多样性保护的典范。

作为国内首个风景名胜区生物多样性保护行动计划，本书在全面梳理景区生物多样性概况、特点、文化价值、价值评价、保护管理现状与进展的基础上，运用分景区空间瞬时计算方法，估算了景区旅游环境容量并对旅游环境的承载能力进行了评价。通过对生物多样性保护功能重要性与保护物种栖息空间的评价与识别，划定生物多样性保护重要区域。结合对生物多样性保护面临问题的深入分析，本书提出了景区生物多样性保护的目标、主要任务，系统规划了 10 个优先领域和 30 项优先行动。

感谢生态环境部环境规划院陆军书记、王金南院士、王夏晖副总工程师，中国环境科学研究院张风春教授、李俊生教授，中国科学院地理科学与资源研究所闵庆文研究员、钟林生研究员，中央民族大学薛达元教授，北京林业大学王建中教授、王丰俊教授，北京师范大学曾维华教授等知名专家学者在成书过程中给予的宝贵建议与帮助。

感谢黄山风景名胜区管委会园林局给予的大力支持和帮助。

感谢参考文献中的所有作者，他们的真知灼见给予本书写作极大的启示，书中引用内容在标注处若有遗漏，谨此致歉。

感谢中国环境出版集团对本书出版给予的大力支持。

本书适合从事风景名胜区生物多样性保护和管理等领域研究的专家学者、相关政府部门的管理者与决策者、生物多样性相关领域的研究生以及其他对此有兴趣的人士阅读参考。若有不当或错误之处，敬请各位专家、学者和广大读者提出宝贵意见！

作　者
2020 年 7 月于北京

目 录

1 绪论 1
   1.1 黄山风景名胜区概况 /1
   1.2 黄山风景名胜区生物多样性保护行动计划的编制由来 /1
   1.3 与《中国生物多样性保护战略与行动计划（2011—2030 年）》的呼应和契合 /2

2 黄山风景名胜区生物多样性保护行动计划基础研究 5
   2.1 黄山风景名胜区生物多样性概况 /5
   2.2 黄山风景名胜区生物多样性的特点 /9
   2.3 黄山风景名胜区生物多样性的文化价值 /11
   2.4 黄山风景名胜区生物多样性价值评价 /12
   2.5 黄山风景名胜区生物多样性保护与利用 /19

3 黄山风景名胜区旅游环境容量与承载力评价 21
   3.1 黄山风景名胜区旅游发展与环境承载力的制约 /21
   3.2 黄山风景名胜区旅游环境容量估算 /22
   3.3 黄山风景名胜区旅游环境承载与评价 /28

4 黄山风景名胜区生物多样性保护空间分区 31
   4.1 黄山风景名胜区生物多样性保护目标分析 /31
   4.2 黄山风景名胜区生物多样性保护重要区域划定 /32
   4.3 黄山风景名胜区生物多样性保护分区管控措施 /41

5 黄山风景名胜区生物多样性保护面临的问题 43
5.1 环境承载力分布不均，自然生态面临瞬时压力 / 43
5.2 生物多样性管护任务艰巨，监测预警能力薄弱 / 44
5.3 资源可持续利用尚未实现，社区集聚发展效率不高 / 45
5.4 地域性传统文化缺乏传承，品牌价值效应尚未显现 / 46

6 黄山风景名胜区生物多样性保护行动计划 47
6.1 保护目标与主要任务 / 47
6.2 黄山风景名胜区生物多样性保护优先行动 / 49
6.3 保障措施 / 54

附　表 57

参考文献 65

CONTENTS

1 Introduction 74
    1.1   Overview of the Mount Huangshan Scenic Area   / 74
    1.2   The Origin of BSAP in the Huangshan Scenic Area   / 75
    1.3   The Relationship between Mount Huangshan and NBSAP   / 76

2 Basic BSAP Research on the Mount Huangshan Scenic Spot 79
    2.1   Biodiversity of the Huangshan Scenic Area   / 79
    2.2   Characteristics of the Biodiversity in the Huangshan Scenic Area   / 85
    2.3   Cultural Value of Biodiversity in Huangshan Scenic Area   / 87
    2.4   Evaluating the Value of the Biodiversity in the Huangshan Scenic Spot   / 88
    2.5   Conservation and Utilization of the Biodiversity in the Huangshan Scenic Area   / 99

3 Evaluation of the Environmental Capacity and Bearing Capacity for Tourism in the Huangshan Scenic Area 102
    3.1   The Restriction of Tourism Development and Environmental Bearing Capacity in the Huangshan Scenic Area   / 102
    3.2   Estimation of the Environmental Capacity for Tourism in the Huangshan Scenic Area   / 104
    3.3   Carrying Capacity and Evaluation of the Tourism Environment in the Huangshan Scenic Area   / 113

4 The Spatial Division of Biodiversity Protection within the Huangshan Scenic Area 117
 4.1 Target Analysis of Biodiversity Protection within the Huangshan Scenic Area　/ 117
 4.2 Delimitation of Important Areas for Biodiversity Conservation within the Huangshan Scenic Area　/ 118
 4.3 Management and Control Measures for Biodiversity Conservation within the Huangshan Scenic Area　/ 131

5 Problems of Biodiversity Protection in the Huangshan Scenic Area 133
 5.1 Uneven Distribution of the Environmental Bearing Capacity of Scenic Spots，a Natural Ecology under Immediate Pressure　/ 133
 5.2 The Tough Task of Biodiversity Management and Protection，Weak Monitoring and Early-Warning Capacity　/ 134
 5.3 The Unrealized Sustainable Utilization of Resources，the Low Efficiency of Agglomerated Community Development　/ 136
 5.4 Lack of a Traditional Regional Culture and Brand Value Effect　/ 137

6 Action Plan for Biodiversity Conservation within the Huangshan Scenic Area 138
 6.1 Protection Objectives and Main Tasks　/ 138
 6.2 Priority Actions for Biodiversity Conservation within the Huangshan Scenic Area　/ 142
 6.3 Safeguard Measures　/ 151

1 绪 论

1.1 黄山风景名胜区概况

黄山风景名胜区位于安徽省黄山市，坐落在黄山区行政区域内（附图1），1982年被国务院列为首批国家级重点风景名胜区，同时也是全国文明风景旅游区和国家5A级旅游景区。该景区1990年被列入世界文化与自然遗产名录，2004年成为世界地质公园，2018年加入世界生物圈保护区网络。

1988年12月，黄山风景名胜区管理委员会成立，现有16个内设机构，其中正县级机构8个：纪工委（监察室、经济审计局）、办公室、政治处、园林局、规划土地处（黄山地质公园管理局）、经济发展局（财政局）、综合执法局（综治办、黄山市旅游管理综合执法局黄山风景名胜区分局）、公安局；副县级机构6个：机关党委、工会、共青团（归口政治处管理）、宣传部（文明办）、人武部、市场监督管理局；正科级机构2个：交通局、花山谜窟—渐江风景名胜区管理处。管委会主任由市长兼任，管委会在黄山市人民政府领导下，负责风景名胜区的保护、利用和统一管理工作。

黄山风景名胜区以旅游为支柱产业，近十年旅游营业收入变化趋势明显，2017年黄山风景名胜区共接待游客336.87万人次，比十年前接待量（2008年224万人次）增长了50.4%。与黄山风景名胜区相邻的黄山区汤口镇、谭家桥镇、三口镇、耿城镇、焦村镇和国有洋湖林场，经济发展状况很不均衡，"五镇一场"的居民收入主要来源于农业、林业和旅游业。

1.2 黄山风景名胜区生物多样性保护行动计划的编制由来

黄山因汇集了奇峰异石、古树名松、云海温泉、冰川遗迹等奇特的自然景观和人文景观资源及丰富的生物多样性而成为闻名天下的风景名胜区，被世界自然保护联盟确定为世界108个生物多样性分布中心之一，被认定为中国33个生物多样性保护优先区域之一（黄山—怀玉山生物多样性保护优先区域），具体范围见图1-1。作为钱塘江重要的周缘山地水源地，黄山及周边社区的生物多样性将直接影响下游钱塘江、淡水湖千岛湖的水质及浙江人民的福祉。黄山的生物多样性质量关系到区域生态安全和经济效益。

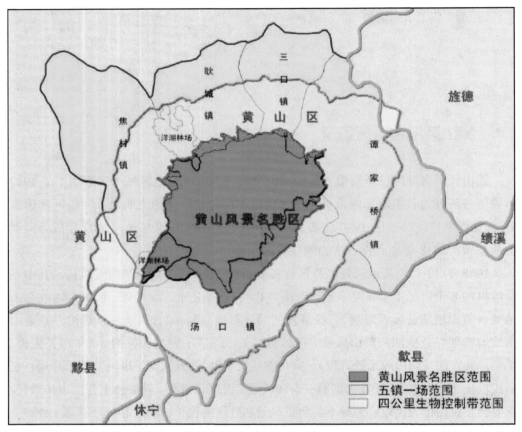

附图1 《黄山风景名胜区生物多样性保护行动计划（2018—2030年）》范围示意

作为国内首个编制生物多样性保护行动计划的风景名胜区，黄山风景名胜区通过科学部署和系统实施生物多样性保护相关行动，能够进一步凸显黄山风景名胜区的生物多样性重要地位和综合效益，不断提升公众教育水平与游览体验舒适度，增加社区幸福度，保证公平受益。

1.3 与《中国生物多样性保护战略与行动计划（2011—2030年）》的呼应和契合

《中国生物多样性保护战略与行动计划（2011—2030年）》（NBSAP）经国务院

第 126 次常务会议审议通过，并于 2010 年 9 月 17 日发布，成为我国生物多样性保护的总纲。《黄山风景名胜区生物多样性保护行动计划（2018—2030 年）》在充分借鉴 NBSAP 的基础上，结合景区实际进行了呼应和创新。

1.3.1 总体目标的契合

NBSAP 确立了贯彻落实科学发展观，统筹生物多样性保护与经济发展的原则，以实现保护和可持续利用生物多样性、公平合理分享利用遗传资源产生的惠益为目标，以 2015 年、2020 年、2030 年为时间节点明确了全国生物多样性保护的阶段性目标。

《黄山风景名胜区生物多样性保护行动计划（2018—2030 年）》以确保景区生态系统服务水平不降低、受保护空间面积不减少为宗旨，严守生物多样性保护底线，以生物多样性保护优先行动为抓手，促进生物多样性保护与景区经济社会发展的协调统一，与 NBSAP 总体目标一致。

1.3.2 重要区域划分思路的呼应

NBSAP 根据我国自然条件、社会经济状况、自然资源及主要保护对象分布特点，将全国划分为 8 个自然区域（即东北山地平原区、蒙新高原荒漠区、华北平原黄土高原区、青藏高原高寒区、西南高山峡谷区、中南西部山地丘陵区、华东华中丘陵平原区和华南低山丘陵区），进而综合考虑生态系统类型的代表性、特有程度、特殊生态功能及物种丰富度、珍稀濒危程度、受威胁程度、地区代表性、经济用途、科学研究价值、分布数据获得度等因素，划定了 35 个生物多样性保护优先区域。黄山风景名胜区属于华东华中丘陵平原区，位于黄山—怀玉山区生物多样性保护优先区域。

《黄山风景名胜区生物多样性保护行动计划（2018—2030 年）》以景区生物生态系统的完整性、特殊物种栖息地和重要遗传资源分布地作为主要参考因素，对黄山生物多样性保护的重要区域进行两个层次的评价与识别，划分为生物多样性保护屏障区、保护物种高适应性区、景观视域区和自然灾害脆弱区四类重要区域。

1.3.3 优先领域与优先行动的呼应

NBSAP 确定了到 2030 年我国生物多样性保护的 10 个优先领域和 30 个优先行动，《黄山风景名胜区生物多样性保护行动计划（2018—2030 年）》也同样部署了 10 个优先领域及 30 个优先行动。

其中，加强生物多样性的调查监测、科学开展生物多样性就地与迁地保护、促进生物资源可持续开发利用、加强外来入侵物种管理、提高应对气候变化的能力、完善生物多样性保护相关政策和制度、推动公众宣传教育等优先领域成为 NBSAP

和《黄山风景名胜区生物多样性保护行动计划（2018—2030年）》共同关注的目标。

此外，根据黄山风景名胜区的基础能力、生物多样性特点和发展需求，《黄山风景名胜区生物多样性保护行动计划（2018—2030年）》在NBSAP的基础上，提出了生物多样性信息化管理、珍稀濒危特有物种的特殊保护、生态系统修复、推动景区周边社区协同发展等优先领域，设置了相应的13项优先行动。

2 黄山风景名胜区生物多样性保护行动计划基础研究

2.1 黄山风景名胜区生物多样性概况

黄山风景名胜区位于中亚热带北缘，地理位置独特，群山起伏，地貌多样，海拔高差1 600多米，是第四纪冰期动植物的避难场，生物多样性丰富，有2 385种高等植物和417种脊椎动物，在全国0.044%的陆地面积上分布着全国6.92%的植物物种和9.55%的动物物种。

2.1.1 生态系统多样性

黄山风景名胜区森林生态系统面积最大，约占景区总面积的89.48%；其他生态系统（如裸露地等）面积其次，约占景区总面积的6.50%，集中分布在温泉、云谷和钓桥管理区；草地生态系统约占景区总面积的0.40%，主要分布在浮溪管理区；湿地生态系统和城镇生态系统面积小，不足景区面积的0.2%，仅零星分布在景区的东南侧部分区域（图2-1）。

森林资源丰富，森林覆盖率约为98.29%，根据遥感数据解译，景区内森林生态系统总面积约为143.7 km^2，比例约为89.48%。从森林类型看，森林生态系统主要包括：以黄山松为建群种的常绿针叶林（120.74 km^2），以壳斗科、樟科乔木为建群种的常绿阔叶林（3.37 km^2），落叶阔叶林（9.42 km^2）和落叶阔叶灌木林（9.30 km^2）等（图2-2）。

2.1.2 植物资源多样性

黄山植物种类繁多，群落复杂多样，自然分布维管束植物178科776属1 924种（包括变种、亚种和变型），如表2-1所示。其中，蕨类植物37科66属161种；裸子植物6科22属39种；被子植物135科688属1 724种。在被子植物中，双子叶植物有 119 科 555 属 1 455 种，单子叶植物有 16 科 133 属 269 种[①]。

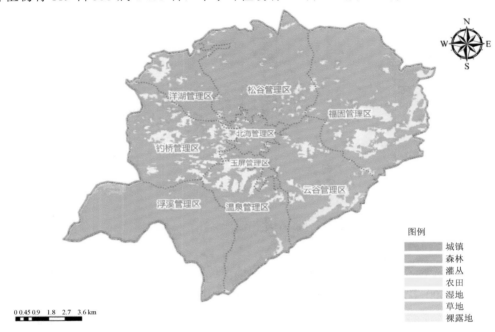

图 2-1　黄山风景名胜区生态系统类型示意

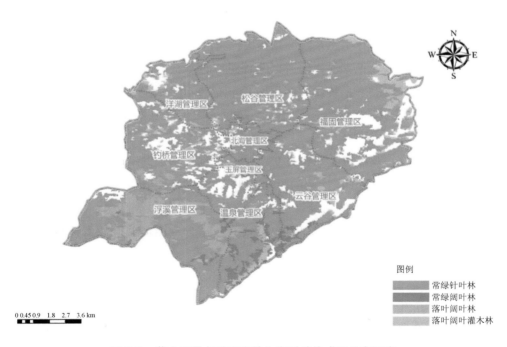

图 2-2　黄山风景名胜区森林生态系统构成及分布示意

---

① 王祥荣，钱阳平. 黄山风景区资源环境评估与基础数据库建构[M]. 北京：科学出版社，2018.

表 2-1　黄山风景名胜区维管束植物成分统计

植物类型	科		属		种	
	数量	比例/%	数量	比例/%	数量	比例/%
蕨类植物	37	20.79	66	8.51	161	8.37
裸子植物	6	3.37	22	2.83	39	2.03
被子植物	135	75.84	688	88.66	1 724	89.60
合　计	178	100	776	100	1 924	100

对种子植物的科、属、种的地理成分分析结果显示：科以泛热带成分为主（占48.5%），其中科的热带性地理成分占59.6%；属以北温带分布最多（占23.7%），其中属的温带性地理成分占61.7%；种以中国特有分布最多，而种的温带性地理成分占33.1%（专栏2-1）。

> **专栏 2-1　黄山风景名胜区植物多样性概况**
>
> 根据吴征镒的《中国种子植物属的分布区类型》，黄山风景名胜区的种子植物有14个分布区类型，只缺少中亚分布类型。世界分布类型有55属，占总属数的8.40%，含10种以上的属有蓼属、铁线莲属、悬钩子属、堇菜属、珍珠菜属和早熟禾属。热带性成分共209属，占总属数的31.91%，其中以泛热带分布及其变型最多，有99属，占热带性成分总属数的47.37%。温带性成分所含属数共362属，占总属数的55.27%，其中以北温带分布类型占有较大的优势，有137属，占温带性成分总属数的37.85%，北温带分布型中栎属、槭树属、荚蒾属等是该区森林植被中的常见种类；其次为东亚分布型，有109属，占温带性成分总属数的30.11%。中国特有分布型有22属，只有腊梅属含2种，其余各属均含1种，其中有许多单种属植物，如明党参属、杜仲属。

2.1.3　动物资源多样性

黄山峰壑纵横，植被茂密，是野生动物理想的繁衍栖息场所，因此，野生动物资源十分丰富。黄山野生动物以东洋界种（即起源于热带、亚热带的种类）占主导地位，其中分布有兽类8目22科73种，占安徽省兽类总数96种的76%，鸟类有17目55科244种，约占安徽鸟类总物种数的60%；两栖类有2目8科29种，爬行类有2目9科54种[①]。

[①] 王祥荣，钱阳平. 黄山风景区资源环境评估与基础数据库建构[M]. 北京：科学出版社，2018.

2.1.4 遗传资源多样性

（1）林木遗传资源种类多样

黄山森林覆盖率高，林木储量丰富，森林类型多样，林木资源丰富，林木遗传资源主要包括黄山松、柏木、鹅掌楸、三角枫、猴樟和毛竹等，见表2-2。

表2-2　黄山风景名胜区代表性林木遗传资源一览

资源种类	种名	科名
林木遗传资源	柏木（*Cupressus funebris*）	柏科
	黄山松（*Pinus taiwanensis*）	松科
	鹅掌楸（*Liriodendron chinense*）	木兰科
	三角枫（*Acer buergerianum*）	槭树科
	猴樟（*Cinnamomum bodinieri*）	樟科
	毛竹（*Phyllostachys heterocycle* cv. *pubescens*）	禾本科

（2）药用植物遗传资源丰富

黄山植物种类繁多，药用遗传资源丰富。据调查，黄山药用植物达700余种，约占植物种数的38.9%，重要药用遗传资源包括八角莲、三叶木通、凹叶厚朴等，见表2-3。

表2-3　黄山风景名胜区代表性药用植物遗传资源一览

资源种类	种名	科名
药用遗传资源	八角莲（*Dysosma versipellis*）	小檗科
	黄连（*Coptis chinensis*）	毛茛科
	三叶木通（*Akebia trifoliata*）	木通科
	凹叶厚朴（*Magnilia officinalis*）	木兰科
	白术（*Atractylodes macrocephala*）	菊科
	明党参（*Changium smyrniodies*）	桔梗科

（3）花卉遗传资源独特

黄山风景优美，奇峰、古树还包括丰富的花卉遗传资源，黄山独特的地理地质条件与植物区系，形成了黄山独特的花卉组成，包括黄山梅、黄山杜鹃、天女花、吊钟花、四照花等，见表2-4。

表 2-4　黄山风景名胜区花卉遗传资源一览

资源种类	种名	科名
花卉遗传资源	黄山梅（*Kirengeshoma palmata*）	虎耳草科
	天女花（*Magnolia siebodii*）	木兰科
	黄山杜鹃（*Rhododendron anwheiense*）	杜鹃花科
	短穗竹（*Brachystachyum densiflorum*）	禾本科
	吊钟花（*Enkianthus quinque*）	杜鹃花科
	四照花（*Dendrobenthamia japonica* var.*chinensis*）	山茱萸科
	春兰（*Cymbidium goeringii*）	兰科
	金樱子（*Rosa laevigata*）	蔷薇科

（4）农作物遗传资源

黄山农作物遗传资源丰富，作为我国著名的茶叶产区，名茶众多，包括黄山毛峰、太平猴魁、祁门红茶、屯溪绿茶等。地区特产茶类"黄山种"（*Camellia sinensis* 'Huangshanzhong'），属山茶科山茶属茶种，原产于安徽省歙县黄山一带，其抗逆性强，产量高，被广泛种植（专栏 2-2）。

专栏 2-2　黄山茶叶资源概况

黄山市已获地理标志的产品有祁门红茶（祁山牌）、黄山毛峰（漕溪牌），原产地域保护的产品有太平猴魁（黄山区三口镇）。2005 年国家质量监督检验检疫总局正式批准黄山区新明猴村茶场、黄山中明茶叶实业有限公司等 6 家企业使用"太平猴魁茶"原产地域产品专用标志，并已由黄山区政府为这 6 家企业正式授牌，确保了黄山市名优茶的产品质量，遏制了假冒产品横行的势头，大大提升了产品的形象和企业的品牌资产价值。

2.2　黄山风景名胜区生物多样性的特点

2.2.1　生态系统类型多样且脆弱

黄山风景名胜区处于亚热带向温带过渡的交汇地带，是长江和钱塘江两大水系在皖南地区的分水岭，处于我国东部中亚热带常绿阔叶林北部地段。

景区内分布有 10 个生态系统类型、16 个主要植被群系（附表 1），随海拔高度变化，其地貌类型自上而下分为起伏平缓的山顶面区、花岗岩强烈切割区、山麓周缘的低山岳陡区，特殊的地质构造与地形地貌导致植被恢复和演化过程缓慢，一旦破坏极难恢复。

2.2.2 特殊生境与群落分布密集

黄山经历了漫长的造山运动和地壳抬升，以及冰川和自然风化作用，形成了特有的峰林结构。块状峰林，以朱砂、眉毛、鳌鱼等峰为代表。筒状峰林，以莲花、天都和莲蕊诸峰为代表，北坡狮子林至芙蓉岭，分布着数条南北方向平行排列的山脊，脊薄如刀刃，称之为刀刃式岭脊；西海排云亭深谷两侧则多是陡悬破碎的峰林。

峰林垂直景观，是黄山最具特色的特殊生境，土壤稀薄，植被难以附着生长。垂直地带主要分布了黄山松、吊钟花等植物，形成了黄山特有的奇松怪石景观。

2.2.3 珍稀濒危保护物种丰富

黄山风景名胜区是地区特有种的集中分布区，以"黄山"命名的特有植物如黄山杜鹃（*Rhododendron anhweiense*）、黄山梅（*Kirengeshoma palmata*）、黄山花楸（*Sorbus amabilis*）、黄山木兰（*Magnolia cylindrica*）等多达 31 种；景区同时分布有国家重点保护野生植物 21 种（附表 2），其中国家Ⅰ级保护植物有 3 种，国家Ⅱ级保护植物有 18 种；列入《中国物种红色名录》的种类有 72 种（附表 3）；拥有《濒危野生动植物种国际贸易公约》（CITES）附录Ⅱ中所列的保护植物 26 种，隶属于兰科（附表 4）；列入国家重点保护的古树名木有 136 株，其中有 54 株被列入《世界遗产名录》；拥有国家保护动物 26 种，其中一级保护动物 8 种，二级保护动物 18 种（附表 5）。这些珍稀濒危的动植物物种面临生态系统脆弱、景区旅游产业发展迅速的客观情况，亟待得到更为妥善的保护。

2.2.4 遗传资源繁多且独特

黄山风景名胜区地域辽阔，生态环境优越，拥有丰富的生物物种资源，蕴藏了大量珍贵的遗传基因多样性，周边地区形成了茶产业、竹木业、茧丝绸业、果蔬业、中药材业和养殖业等六大主导产业和特色产品。

黄山保存有多种国家珍稀濒危药用植物资源，具有重要的经济价值、科研价值和药用价值，如野大豆对培育或改良新品种具有重要意义，金钱松、鹅掌楸、黄山梅等在亚植物区系和植物科的系统发育和分类等方面有重要意义。黄山因其独特的生态环境，保留了第四纪冰川时期中的许多古老的珍稀树种，如红豆杉、连香树等，是我国独具特色的绿色宝库。

2.3 黄山风景名胜区生物多样性的文化价值

2.3.1 徽州文化体系博大精深

黄山地处文化历史悠久的古徽州，历经约 800 年的衍变过程，形成了以"和""善""儒"为核心理念，包含观念文化、制度文化和地域乡土文化三大内涵的徽州文化体系，具有鲜明的地域特色。以徽州文化为内核，衍生出以古代建筑、古村落为载体的物态文化，以民间艺术、民俗文化为依附的非物态文化，新安理学、徽州新学、新安画派、新安医学等文化流派及贾而好儒的徽商文化。

2.3.2 传统风俗技艺极具价值

徽州文化中婚嫁习俗、上九庙会、渔梁跳钟馗、叠罗汉、抛绣球、傩舞、目连戏等徽州民间风俗和技艺独具地方特色；传统民间艺术纷繁多样，徽州戏曲、"徽州四雕"、篆刻、版画极具历史和艺术价值。

2.3.3 文艺宗教瑰宝内涵丰富

黄山现存赞美黄山的诗词达两万余首，"黄山画派"自明末清初流传至今，以黄山为题材的书法、绘画、摄影作品不计其数；景区有慈光阁、半山寺、云谷寺、松谷庵、翠微寺等道教、佛教的历史文化遗存，保存古道、古桥、古寺、古亭等古建筑近百处，摩崖石刻 200 多处，是黄山人文景观不可或缺的部分。

2.3.4 民居建筑风格彰显特色

黄山地区的民居、祠堂和牌坊并称"古建三绝"，最具古徽州特色，现已发现的地面文物古迹达 5 000 多处，其中列入世界文化遗产的有黄山市的黟县西递、宏村古村落 2 处，列入国家和省级重点保护单位的达 61 处。

2.3.5 自然人文结晶水乳交融

徽州文化与黄山的气候物产资源交汇，自然环境与人文环境相互浸润，形成了黄山特色的区域文化结晶。黄山茶与徽州茶道相辅相成，以物产为原料、以工艺为传统的徽州文房四宝驰名中外；可供食用的菌类、笋类有 800 多种，兽类有 80 余种，淡水鱼类丰富多样，为徽菜的形成和发展提供了良好的物质基础。

2.4 黄山风景名胜区生物多样性价值评价

2.4.1 生物多样性直接开发价值评价

直接开发价值是指所拥有的生物多样性及其构成的生态资产所产生的最为直观的开发利用产生的市场经济价值[①]，主要包括开展休闲游憩、科学研究、文化教育等活动产生的经济效益。

将黄山风景名胜区的直接开发价值划分为旅游收入、科研价值、文化价值和产品价值四个部分，经过初步估算，黄山风景名胜区的直接开发价值约为 35.12×10^8 元/a，其中旅游收入产生的直接价值所占比例达到 80%以上，见图 2-3。

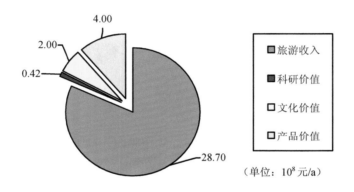

图 2-3 黄山风景名胜区的直接开发价值估算及构成示意

2.4.2 生物多样性生态保护价值评价

生态保护价值是指虽然没有得到货币的直观体现，却为旅游生产和消费的正常进行提供了必要的自然支持，也为区域的生态环境提供调节和支撑服务的价值，包括了生态效益和生物资源两部分价值。经初步估算，黄山风景名胜区生态保护价值约为 208.54×10^8 元/a，其中生物资源价值量占绝对比例（95%左右）。

（1）生态效益价值

生态效益价值主要是指自然生态系统所能够提供的涵养水源、保持土壤、调节气候、控制灾害和净化环境的功能，通过公式核算方法对各部分功能进行货币价值的衡量，得出黄山风景名胜区生态效益价值约为 11.16×10^8 元/a，功能构成及估算

[①] 丁晖，徐海根. 生物物种资源的保护和利用价值评估——以江苏省为例[J]. 生态与农村环境学报，2010，26（5）：454-460.

结果见图2-4。

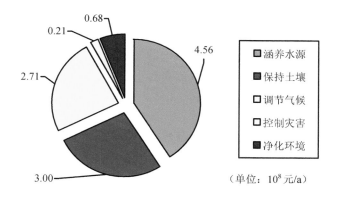

图 2-4 黄山风景名胜区的生态效益价值估算及构成示意

1）涵养水源功能

涵养水源功能主要包括了生态系统对水资源的渗透能力和蓄水能力，其价值主要体现在增加有效水量、改善水质和调节径流上，通常利用市场定价法进行估算，即

涵养水源价值（元/a）=年涵养水源总量（t）×当年居民用水价格（元/t）

注：通常假定年涵养水源总量=年径流量=年平均降水量－年平均蒸腾量

根据黄山风景名胜区的水文水利相关数据，景区年平均涵养水源总量约为 $233.67×10^6$ t，农用水价格约为 1.95 元/t，估算景区年涵养水源功能价值量约为 $4.56×10^8$ 元/a。

2）保持土壤功能

土壤是重要的不可再生资源，是经过自然生态系统长久的生物和物理过程积累而成，其价值体现在能够为生物生存发育提供场所，降解有机物质，保持肥力，保障养分供给，促进地球自然循环，是国家和地区财富的重要组成部分。

其中，防止水土流失的价值评估可以采用机会成本法，计算公式如下：

土壤侵蚀总量=单位面积水土流失量差异×生态系统总面积

废弃土地面积=土壤流失总量/表层土壤平均厚度

土壤侵蚀的价值=该生态系统产生的平均收益×相当的废弃土地面积

此外防止土壤养分流失的价值，一般用总氮、总磷、总钾的含量来衡量，可以通过市场定价法计算，其公式如下：

土壤养分流失价值=土壤流失总量×土壤层容重×

土壤中氮磷钾单位量×三种养料的市场价格

据统计资料显示，黄山市水土流失的总量达到250km^2，相当于7.28万t化肥量，土壤养分流失的价值约达 3.00×10^8 元/a。

3）调节气候功能

生态系统中的绿色植物能够固定大气中的二氧化碳，并在生产过程中调节大气氧含量的变化，保证生命活动的基本气候条件。因此，衡量调节气候功能可以通过固碳释氧法进行初步估算，公式如下：

固碳——碳税法：固碳价值=生态系统二氧化碳年吸纳量×影子价格

释氧——替代成本法：释氧价值=总生物量×1.2×工业制氧单位成本

碳氧系数是通过光合作用来确定的：

固定碳量=固定二氧化碳量×0.27

释氧量=植物生产量×1.2

经过初步统计估算，黄山风景名胜区森林植被的碳储量约为 57 742.55 t，折合二氧化碳 213 861.296 t。以 2017 年湖北省碳排放权交易中心的二氧化碳交易价格 19.11 元/t 为影子价格进行估值计算，得出景区碳储量价值为 2.71×10^8 元/a。

4）控制灾害功能

控制灾害功能主要包括防止旱涝、山洪等自然灾害和生物灾害的功能。可以考虑用防止损失成本法估算，公式如下：

防止自然灾害的价值=生态系统总面积×单位面积损失量

控制生物灾害的价值=单位面积防止生物灾害的投入成本×生态系统面积

根据原国家林业局发布的技术文件显示，林地减轻水旱灾害的年效益约为65元/hm^2，黄山风景名胜区林地面积为 12 271.1 hm^2，计算得出景区减轻水旱灾害生态效益为 797 621.5 元/a；通过统计，黄山风景名胜区用于森林防火、森林病虫害等方面的投入大约为 0.20×10^8 元/a。因此，控制灾害功能的价值约为 0.21×10^8 元/a。

5）净化环境功能

净化环境功能指通过物质生态循环进行生态系统的净化。可以使用替代成本法来进行估算，公式如下：

处理废物价值=生态系统总面积×生态系统吸收该污染物的单位量×
工业净化该污染物的单位成本

林地主要发挥了对粉尘、有害气体的吸收净化功能，根据原国家林业局技术文件规定，阔叶林的滞尘能力为 10.11 t/（hm^2·a），削减成本为 170 元/t；吸收二氧化硫量为 88.65 kg/（hm²·a），削减二氧化硫成本为 600 元/t；吸收氮氧化物量为 0.38 t/（hm²·a），削减氮氧化物的成本为 250 元/t，根据黄山风景名胜区林地面积（1.2 万 hm²）初步估算，黄山风景名胜区的净化气体服务功能价值约为 $0.23×10^8$ 元/a。

此外，依据谢高地的生态服务功能价值当量初步估算，黄山风景名胜区的废物处理功能价值约为 $0.45×10^8$ 元/a。景区的净化环境功能价值合计约为 $0.68×10^8$ 元/a，见图 2-5。

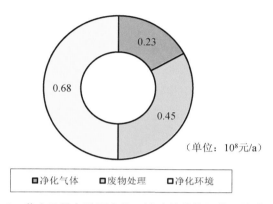

图 2-5　黄山风景名胜区净化环境功能价值估算及构成示意

（2）生物资源价值

生物资源是实现可持续发展战略的重要资源，人们对生物资源利用的实践超前于对其理论上的研究，认识和研究生物资源的价值对部署地区经济发展战略和物种保护十分重要。生物资源本身具有一定价值，有的可以直接进入市场流通，体现经济价值，它的价值就可以通过经济效益有明确的标价；另外一些生物资源为人类提供了源源不断的服务，但难以直接标明价值。目前国际上确定生物资源价值的方法有多种，依各国家的标准和状况而不同，得到公认的有以下三种评价角度：

①评价自然状态下的生物价值。

②评价具有商业性收益的生物产品的价值。

③评价生态系统服务功能中的生物多样性服务价值。

黄山风景名胜区生物资源价值由三部分构成（图 2-6），由于尚未形成明确的生物资源价值估算方法，仅就黄山地区具有特色的代表性生物资源进行价值估算，总价值约为 $197.38×10^8$ 元/a。

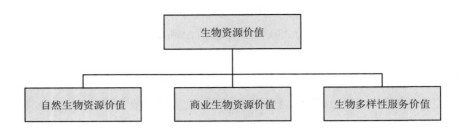

图 2-6　黄山风景名胜区的生物资源价值构成示意

1）自然状态下的生物资源价值

自然生物资源在区域发展和居民生产生活的过程中发挥着极其重要的作用，同时也是景区重要的旅游资源，其中供给食物生产和原材料的价值以谢高地等研究制定的生态服务价值当量因子表为基础进行初步估算，见表 2-5。

表 2-5　黄山风景名胜区自然状态下的生物资源价值估算

类别	林地	草地	农田	湿地	水体	荒地	城镇
食物生产当量	0.1	0.3	1.0	0.3	0.1	0.01	0.00
食物生产单位面积价值/（元/hm²）	201.22	603.65	2 012.15	603.65	201.22	20.12	0.00
食物生产价值/（万元/a）	303.17	0	10.66	0	2.56	0	0
原材料提供当量	2.6	0.05	0.1	0.07	0.01	0.00	0.00
原材料提供单位面积价值/（元/hm²）	5 231.6	100.61	201.215	140.85	20.12	0	0
原材料提供价值/（万元/a）	7 882.45	0	1.07	0	0.26	0	0
自然生物资源价值/（亿元/a）	0.82						

除提供食物生产和原材料的生物资源之外，黄山地区拥有众多代表性的自然生物资源，这些资源的存在也成为黄山风景名胜区在国内同行业中的独特标志，下面介绍两类资源的价值评估作为参考：

a. 古树名木价值：以黄山松、黄山梅、银杏等古树名木为代表，据《徽州古树》记载，黄山古树约有 110 种，最早可追溯至唐代；黄山"四绝"之首的"奇松"是黄山松，也是游客观赏黄山风光的必选景点。

以已有研究资料对九华山古树名木资源价值的评估为基础，参考其主要数据对黄山风景名胜区古树名木资源价值进行估算，见表 2-6。

表 2-6 黄山风景名胜区古树名木资源价值估算

	游客接待人次/万人	支付意愿值 1/（亿元/a）	支付意愿值 2/（亿元/a）	均值合计/（亿元/a）
九华山风景区	401	1.500	4.589	3.04
黄山风景名胜区	299.1	1.119	3.423	2.27

b. 珍稀动植物价值：黄山风景名胜区分布有国家Ⅰ级保护植物 3 种，Ⅱ级保护植物 18 种；国家Ⅰ级保护动物 8 种，其中鸟类 2 种，特有动植物黄山短尾猴、黄山石杉、黄山溲疏等几十种。

2）商业生物资源价值

a. 药用生物资源：黄山市地产药材种类繁多，蕴藏量也很可观。据中药资源普查统计，常用的大宗药材达 400 余种。其中蕴藏量或年产量在 100 t 以上的有白花前胡、防己、土茯苓、合欢皮、大血藤、（贡）菊花、垂盆草、山楂、覆盆子、竹沥等 66 种；蕴藏量或年产量在 10~100 t 的有（徽）白术、首乌、桑白皮、辛夷花、梅花、过路黄、山茱萸、女贞子、薏苡仁、海金沙、五倍子、夜明砂等 131 种。按照市场售价估算药用资源价值大约为 13.8 亿元/a。

b. 黄山茶业资源：根据黄山市统计数据，2013 年黄山春茶产量约为 12 966 t，与上年同期基本持平，均价为 61.8 元/kg，同比增长 25.1%，产值达 80 165 万元。

此外，根据央视财经频道联合中国品牌建设促进会等单位发布的 2015 年中国品牌价值评价信息，区域品牌（地理标志保护）黄山太平猴魁茶，品牌强度为 920.0，区域品牌价值为 111.43 亿元。黄山除太平猴魁外，还有黄山毛峰、祁门红茶等国际知名品牌，以此估算黄山茶品牌价值约为 150 亿元。

c. 野生山珍资源：徽菜是全国八大菜系之一，其中山珍是徽菜的主要特色，如毛竹笋、野山笋、蕨菜、食用菌是景区主要的土特产。

d. 以生物原料为主的区域传统文化资源：工艺品、中药材及其他类地理标志产品，如宣纸品牌强度达到 920.0，区域品牌价值为 21.49 亿元。

3）生物多样性服务价值

黄山风景名胜区动植物资源较为丰富，有着巨大的潜在价值，这部分价值可采用机会成本法或重置成本法进行估算，也可以根据谢高地等研究制定的生态服务价值当量因子表，对生物多样性保护的生态价值进行初步估算，见表 2-7。

表 2-7　黄山风景名胜区生物多样性价值估算

	林地	草地	农田	湿地	水体	荒地	城镇
生物多样性价值当量	3.26	1.09	0.71	2.50	2.49	0.34	0.00
黄山单位面积价值/（元/hm²）	6 559.62	2 193.25	1 428.63	5 030.38	5 010.26	684.13	0.00
黄山各类型生物多样性价值/（万元/a）	9 883.38	0.00	7.57	0.00	63.63	0.00	0.00
黄山生物多样性价值/（亿元/a）	1.00						

2.4.3　生物多样性未来存在价值评价

未来存在价值是指黄山目前没有得到直接开发利用，但是伴随着黄山资源而客观存在的、可以供子孙后代或人们自己将来使用的价值，包括存在价值、遗产价值和选择价值。其中，存在价值是指人们为确保黄山旅游资源永续存在而自愿支付的费用；遗产价值是指当代人将黄山作为旅游资源保留给子孙后代而支付的费用；选择价值是指个人为自己、为自己子孙后代或为别人在将来能够有选择地使用黄山旅游资源而每年预先自愿支付的一笔保险金。对于这些类型的生态系统服务，只能采用假想的市场评价方法，即意愿调查法（CVM）进行评价。

通过对研究文献、统计数据和经济增幅估算，2017 年人均对黄山生态系统保护的最大支付意愿为 27.96 元，按 2017 年全国总人口 139 008 万人，城镇常住人口 81 347 万人，初步估算黄山风景名胜区未来存在价值范围为 $227.45 \times 10^8 \sim 388.67 \times 10^8$ 元/a。

2.4.4　黄山风景名胜区生物多样性价值评价结果

经过估算，黄山风景名胜区生物多样性价值的组成和估算结果如图 2-7 所示，其中，生态保护价值和未来存在价值属于非直接使用价值，尚未得到直接的货币体现，但是其价值估算值要远远高于直接开发所取得的收益，意味着黄山风景名胜区具有非常重要的保护价值。

从可持续发展的角度来说，直接开发价值是当代人利用黄山旅游资源所得到的市场经济效益，生态保护价值体现了黄山风景名胜区对于地区生态环境和生态安全保障的生态效益，未来存在价值体现了我们的后代子孙能够从黄山的存在中所得到的效益。从价值量来说，黄山的保护应重于开发，保护是第一位的，开发是第二位的。

黄山之所以有巨大的使用价值和存在价值，与黄山独特的地质地貌、多样的生物资源、优美的山水风光、深厚的文化底蕴紧密相关。但这些独特的景观很容易在人类不合理活动的干扰下受到破坏，使黄山旅游资源价值降低。

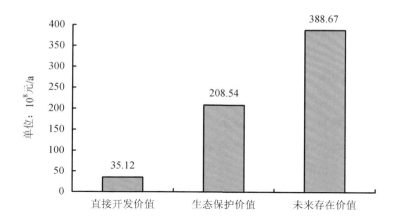

图 2-7　黄山风景名胜区生物多样性经济价值构成示意

做好保护，可以使黄山旅游资源永续存在，直接开发价值、生态保护价值和未来存在价值都得以提升。黄山保护好了，黄山开发也就有了资源依托，开发产生的经济效益就是使用价值的体现，保护得越好，通过开发能实现的价值也会相应提高。同时，开发可以为黄山保护筹集相应的资金，促进黄山的保护，并在将来产生更高的经济价值。

2.5　黄山风景名胜区生物多样性保护与利用

2.5.1　黄山风景名胜区生物多样性保护与管理情况

（1）首创国内景点"轮休"制度

黄山于1987年10月首创了国内景区的景点"轮休"制度，封闭期间对景点自然生态系统、野生动植物资源进行自然和人工修复，每个轮休期为3~5年，莲花峰、狮子峰、丹霞峰、天都峰、始信峰等经典旅游景点已先后开展封闭轮休，中亚向北亚热带过渡性地带的自然植被类型和垂直植被带谱得到有效维续。

（2）加强古树名木保护

黄山风景名胜区登记建档的古树名木已有136株，其中54株被列入世界自然遗产名录。景区对游道两旁的树木包围竹片，定期对古树名木进行全面调查和病虫害防治，邀请专家对古树名木进行体检和会诊，对其采取综合保护措施，对重点古树名木确定专人进行日常养护，编制古树名木管理软件进行信息化管理。

（3）严格开展植物检疫和防火措施

建立以预防松材线虫病为重点的病虫害监测网络，以景区内松林为重点，设置监测站、分站、监测点和检疫检查站，坚持每年春秋两季进行松材线虫病专项普查，并及时清理枯死树木。注重景区防火基础设施、防火宣传、火源监控、协作联防等方面的建设，实现连续39年无森林火灾的佳绩。

（4）积极开展水资源保护

景区相继建成塘库、蓄水池等蓄水设施的满足用水需求，严格管控景区污水和废弃物的排放和末端处理，限制区内宾馆饭店的数量，从根源上保护了水源水质，为众多物种提供了优良的栖息环境，有利于生物多样性的保护，为钱塘江流域及江浙平原水源提供了良好的水质，保障了流域内居民的用水安全。

2.5.2 黄山风景名胜区生物多样性利用概况

自然生态禀赋好，旅游开发效益突出。黄山风景名胜区自然植被保存较完整，聚集程度高，斑块群连通性好，有利于为濒危物种提供完整的栖息场所；岩石景观破碎度较高，为游客提供了多样的石林景观；以奇松、怪石、云海、温泉、冬雪"五绝"著称，现已开发形成温泉景区、云谷景区、玉屏景区、北海景区、天海景区、西海景区、钓桥景区、东海景区等，自1979年开放以来，游客人数不断增加，海外游客接待人次也不断增加，旅游市场日趋成熟。

生物资源开发价值高，为周边社区发展提供助力。黄山林木、中药材、花卉、农作物等生物资源极具发展潜力，景区周边"五镇一场"居民从事旅游服务产业之余，林下种植养殖业、中药材种植销售、原产地茶叶生产、花卉盆景生产等其他特色产业蓬勃发展，成为居民重要的增收途径。

人文景区发展速度加快，与自然资源利用相互影响。景区内人文景点与艺术作品保存较为完整，文学、书法、绘画、摄影等作品层出不穷；黄山市内的屯溪老街、休宁齐云山、黟县西递和宏村、歙县棠樾、徽州古城等一大批自然人文景区旅游事业正蓬勃发展，大黄山景区的综合生态产品初具规模；森林漫步、徒步登山、田园养生、温泉疗养、美食茶品等休闲养生度假发展潜力巨大；徽州文化体系具有鲜明的地域特色，与黄山的气候物产资源交汇，自然环境与人文环境相互浸润，形成了黄山茶、徽州文房四宝、新安医学、新安画派、徽菜等凸显黄山特色的区域文化结晶。

3 黄山风景名胜区旅游环境容量与承载力评价

3.1 黄山风景名胜区旅游发展与环境承载力的制约

3.1.1 黄山风景名胜区旅游产业发展的特点

（1）仍会呈现一段时期内的爆发式持续增长

从国家层面来看，随着居民收入增加、"带薪休假"逐步落实、汽车时代全面来临，民众用于旅游的花费越来越高，旅游消费成为一种刚需，旅游形式将由观光旅游向休闲旅游和度假旅游转变，大众旅游时代将全面来临。从区域尺度来看，黄山风景名胜区地理位置优越，长三角地区和长江中游地区民众自驾即可抵达，航班、高铁的开通带来全国乃至国际范围内更多客源，人数增长趋势可以预计。

（2）以"观光旅游""中短距离旅游"为主

当前黄山风景名胜区的观光旅游仍处于主体地位，享受型、文化型的休闲旅游项目比较少；就旅游的地域性和时间期限而言，一般以中短距离旅游为主，远距离旅游相对较少，旅游时间期限较短，一般多为两三天或三五天，"一日游"也占有很大比重，长期旅游比较少，休闲养生和度假旅游大多以周边城市的老年群体为主，所占比例较小。

（3）智能景区、大数据管理不断融入

黄山风景名胜区将加速智慧化建设作为提升旅游业竞争力的重要方面来强力推进，形成了覆盖全山的信息化网络，在景区资源保护、景区管理、旅游服务、经营发展等方面发挥了积极作用，"码上游黄山"智慧旅游服务平台、客流车流预测、北斗导航等加快了景区的发展。

3.1.2 黄山风景名胜区环境承载力面临的主要压力

（1）资源空间承载压力

资源空间承载压力突出表现在部分景点和通道，如索道的上下站、北海宾馆前、光明顶、鲫鱼背、莲花峰、梦笔生花观景台等，黄山资源空间超载问题早已引起管理人员的关注，并采取了措施，但问题尚未得到彻底解决。

（2）生态环境承载压力

旅游活动对景区的生态环境产生了一些潜移默化的影响，随着时间推移逐步显现，以黄山四绝之首的奇松为例，在人流经常通过的景点或山道两侧常常出现有创伤甚至枯死的松树，这与旅游旺季的人为破坏，人流踩踏造成土壤板结、渗水性差，树木根部受损有直接关联。

（3）供水压力

供水一直是黄山旅游发展的主要环境制约因素之一，由于水源不足、蓄水条件差，游客及工作人员用水常常发生困难，缺水最严重的景区是北海、玉屏和钓桥，缺水量最大的时段是每年9—10月。供水不足除导致游客用水困难以外，还影响到景区的植被保护，自然形成的溪泉水被人工截流输往宾馆，引起植被缺水和水景观美感度降低。

（4）游客心理承载压力

旅游旺季游人漫山遍野，一度出现"拨开人群看风景"的景象，这是全国范围内的旅游景点都存在的共性问题，尤其在国庆节"旅游黄金周"、各小长假等期间，黄山旅游旺季的索道排队、厕所拥挤、观景平台人满为患等问题已经超出游客的心理承载力负荷。

3.2 黄山风景名胜区旅游环境容量估算

3.2.1 空间环境容量

采用分景区空间瞬时计算方法，估算黄山风景名胜区在夏半年旅游旺季（日均开放时间为12 h）景区环境空间上实际所能承受的游客最大限额，景区内主要空间环境可分为线路观光区和平台观光区两部分，分别采用线路法、面积法进行测算，根据《风景名胜区总体规划标准》（GB/T 50298—2018）中关于游客容量的计算标准（表3-1），用游客完成整个景区主要景点的平均游览时间计算日周转率，对主要景点所处景区的空间环境容量进行加和。

表 3-1 黄山风景名胜区空间环境容量计算标准

主要类别	容量测算依据	测算标准
机动车观光线路	按照疏散景区内游客的主要机动车数量和车速、路宽进行测算，设定单车道 1～2 km 一辆载员 25 人的标准旅游车	25m/人
步行观光道、观景平台	遵照《风景名胜区总体规划标准》(GB/T 50298—2018) 中关于容量测算的标准值为 5～8 m^2/人，极限值按照 5 m^2/人计算，非高峰日合理容量按照 8 m^2/人计算	5～8 m^2/人
经典观景点	景区面积小、游客集中的精华观景景点，根据游客心理承受容量，测算标准值为 1～1.5 m^2/人，极限值按照 1 m^2/人计算，非高峰日合理容量按照 1.5 m^2/人计算	1～1.5 m^2/人
生态游线	满足游客野外探险需求，部分景区新增线路，设定平均路宽为 1 m，容量取值为观光道标准值的 5 倍	25～40 m^2/人

以《黄山风景名胜区总体规划（2007—2025 年）》、黄山风景名胜区旅游导览图（图 3-1）、《黄山风景名胜区钓桥景区（扩大）规划（2007—2020 年）》、《黄山风景名胜区东海景区（福固管理区）详细规划》和黄山风景名胜区云谷索道下半段建设项目（索道上站至石笋矼游步道设计方案）为数据基础，对各类型空间环境容量进行测算。

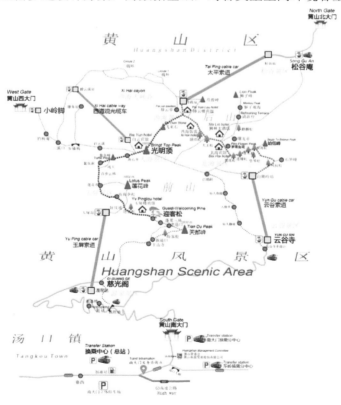

图 3-1 黄山风景名胜区主要旅游线路示意

注：图片来自景区官方网站 http://hsgwh.huangshan.gov.cn/News/show/1043916.html。

(1）步行道环境空间容量测算

统计得到黄山风景名胜区内主要观光步行道长度共约为 74 km（其中新增步行道长度约为 14 km），新增生态游线约为 18 km，取夏半年游览时长为 12 h，游客容量极限值为 5 m^2/人（生态游线 25 m^2/人），规划期内（2018—2030 年，下同）预计景区夏半年的步行道所能容纳的极限游客容量为 53 763 人/d。

（2）机动车观光路环境空间容量测算

统计得到黄山风景名胜区规划期内主要机动车观光路长度约为 29 km（新增机动车线路长度约为 16 km），取夏半年游览时长为 12 h，游客容量极限值为 25 m/人，预计景区夏半年的机动车线路极限游客容量为 3 146 人/d。

（3）观景台、观景点的环境空间容量测算

黄山风景名胜区规划期内主要观光平台（含景点）面积大约为 15 万 m^2，取夏半年游览时长为 12 h，游客容量极限值为 5 m^2/人（经典观景点取 1 m^2/人），景区夏半年观景平台（点）所能容纳的极限游客容量为 7 936 人/d。

（4）各景区空间环境容量及卡口比较

目前，黄山风景名胜区多数游客选择南大门（汤口）作为主要的进出点，观光车、运输车次多，温泉景区和云谷景区是大多数游客参观的第一个游览区，从空间关系上来说，温泉、云谷景区是到黄山精华游览区的入口或引景，游客集中空间固定，停留时间较短，日周转率高；随着北大门索道、西海观光缆车的陆续开通，开始分流到北大门、西大门，因此钓桥景区、东海景区将主要起游客的分流、延伸作用，游客在此停留时间较短，人均享有面积较大，空间容量也较大。

玉屏、北海、天海、西海四个景区是黄山风景名胜区的核心部分，天都峰、莲花峰、光明顶、始信峰等经典景点的游览率几近 100%，天海、北海景区同时属于游客的餐饮休憩、转换游路、人流集散等非常集中的地点，游客停留时间长，这四个景区在旅游旺季会成为影响景区旅游安全和游客容量的卡口。

初步估算，规划期内玉屏景区旺季游客日极限容量为 7 061 人次，北海景区旺季游客日极限容量为 6 110 人次，天海景区旺季游客日极限容量为 9 579 人次，西海景区旺季游客日极限容量为 5 254 人次；黄山风景名胜区旅游旺季的游客日极限容量为 64 846 人次，见表 3-2。

（5）规划景区设施容量

景区内共有玉屏、云谷、太平三条索道和西海观光缆车四条便捷线路，主要承担游客的上下山运载，从索道运力和周转时长来估算，每日索道的最大运力约为 11 万人次，可以满足景区游客运载总体需求。不可避免的是，从核心景区的瞬时游客容量来看（表 3-3），核心景区会在旅游旺季时，因瞬时游客量大造成等待索道的游客积压的情况，见表 3-4。

表 3-2　黄山风景名胜区旺季游客日环境空间容量　　　　　　　　　　　　单位：人次/d

	温泉景区	云谷景区	玉屏景区	北海景区	天海景区	西海景区	钓桥景区	东海景区	合计 2
步行道容量	955	4 631	6 335	3 798	8 367	3 784	14 664	11 230	53 764
机动车线路容量	1 140	—	—	—	—	—	1 526	480	3 146
平台景点容量	—	664	726	2 312	1 212	1 470	1 112	440	7 936
合计 1	2 095	5 295	7 061	6 110	9 579	5 254	17 302	12 150	64 846

表 3-3　黄山风景名胜区核心景区瞬时游客容量　　　　　　　　　　　　单位：人

	玉屏景区	北海景区	天海景区	西海景区	合计 4
线路瞬时容量	2 371	1 597	4 183	1 468	9 619
观景台瞬时容量	190	862	534	450	2 036
经典景点瞬时容量	90	270	—	40	400
合计 1	2 651	2 729	4 717	1 958	12 055

表 3-4　黄山风景名胜区索道设施容量及极限等待时长估算

	最大单向运量/（人/h）	运行时长/h	最大运输能力/（人/d）	单次瞬时游客极值/人	预计等待时长/h
玉屏索道	2 400	10	48 000	12 055	5.0
云谷索道	2 000	10	40 000	12 055	6.0
太平索道	600	10	12 000	12 055	—
西海观光缆车	800	10	16 000	12 055	—
合计 5	5 800	—	116 000	—	2.0

3.2.2　生态环境容量

旅游地的生态环境承载力取决于三个变量，一是旅游地自然生态系统净化与吸收污染物的能力，二是旅游地人工系统处理污染物的能力，三是单位时间内人均产生的污染物数量，包括游客和景区管理人员、服务人员等常住人口。

相较于自然净化能力，人工系统的净化能力和游客及常住人口产生的污染物数量的可控性更强，因此计算生态环境承载力时更多参照人工系统的污染物处理能力（表 3-5）。

表 3-5 黄山风景名胜区生态环境容量测算

评价项目	公式	参数值	数据说明	容量估算
大气环境容量	$B_1 = \dfrac{S \times f}{S_m}$	$S=160.6 \text{ km}^2$ $f=98.29\%$ $S_m=1.7\times10^{-7}$ km²/人	S_m 通过人均 CO_2 排放量和转换率获得，人均 CO_2 排放量约 0.9 kg，CO_2 林地吸纳平均系数 5.2 t/hm²	9.3×10^8
污水处理环境容量	$B_2 = \dfrac{X}{Y}$	$X=4.0\times10^6$ m³/d $Y=0.082$ m³/（人·d）	污水处理量按规划值 166 t/h 估算，Y 取平均值	4.9×10^7
固体废物处理环境容量	$B_3 = \dfrac{W}{Z}$	$W=2.1\times10^4$ kg $Z=0.3$ kg/d	实施"统一标准、规范操作，统一运输、集中处置，统一价格、费用分摊，统一制作、密闭封装"，生活垃圾全面下山	7.0×10^4

根据表 3-5 测算结果，可以得出：

日生态环境容量 $B=\min\{9.3\times10^8, 4.9\times10^7, 7\times10^4\}=70\,000$ 人

需要补充说明的是，基于垃圾处理能力的生态容量计算是以垃圾日产日清为前提的，实际上游客分布具有季节不均衡性，在旅游高峰日，可以增加垃圾承运人员、加大索道运载力度等，因此，固体废物的承载量可以适当高于计算结果，如果时变系数取 2，生态环境容量可达 14 万人/d。

3.2.3 经济环境容量

衡量黄山风景名胜区的经济环境容量主要考虑住宿、供水和交通设施容量三个部分。其中，住宿设施分为景区内和景区外两部分，为了更好地保护景区生态资源，内部的宾馆和餐饮场所正在逐步减少，景区外的汤口、寨西、山岔等处分布越来越多的个体经营中低档床位，加之黄山游线的不断拓宽，屯溪、歙县等大环线上的宾馆、民宿也在旅游旺季承接了大量游客，景区外住宿设施流变性较大且通常起到缓解景区内旅游旺季住宿困难的作用，故不计在景区住宿设施环境容量的考量范围之内。仅估算景区供水、交通设施容量。

（1）景区供水设施容量测算

黄山风景名胜区已建成 8 座塘库，总蓄水量为 243 860 m³；20 个蓄水池，总蓄水量为 5 880 m³。此外有温泉配水站，年可供水量为 132 000 m³，供水设施总蓄水量为 381 740 m³。按枯水年计算，保证率为 95%，复蓄系数取 3.5，年供水利用率取 90%，则黄山风景名胜区每天提供用水为

$$（381\ 740×95\%×3.5×90\%）÷365=3\ 129.75\ m^3/d$$

选用《风景名胜区总体规划标准》（GB/T 50298—2018），人均用水量：散客人均用水为 0.02 m^3/d，宾馆用水标准取每个床位平均值为 0.2 m^3/d。景区内有床位 5 367 个，入住率为 60%，宾馆住宿游客耗水量为 644 m^3/d。山上职工每日用水量上限约为 0.15 $m^3/$（人·d）×865 人=129.75 m^3/d。供水设施所能提供的游客容量为

$$（3\ 129.75 - 644 - 129.75）÷0.02 = 117\ 800\ 人$$

（2）景区交通设施容量测算

据《2017 年黄山风景名胜区统计年鉴》，景区共有旅游车辆 155 辆，其中大型客车 141 辆，中型客车 10 辆，小型车 4 辆。假设各类车辆载客量分别以 50、40、20 人计，班车按照 20 min 发出一次，主要分为"景区换乘中心—云谷寺"和"景区换乘中心—慈光阁"两条主要线路，日单向运输能力约为 29 万人，见表 3-6。

表 3-6 黄山风景名胜区交通设施容量测算

车型与单次核定人数/人		单向发车频次/h	运行时长（夏季）/h	交通设施容量/人
大型车	141×50			
中型车	10×40	7 530	3	13	293 670
小型车	4×20			

3.2.4 社会心理环境容量

（1）当地居民心理容量

黄山风景名胜区已经实现了居民区与风景区分离，景区的常住人口为直接从事旅游业的服务人员和管理人员，这部分居民的心理承载力通常近于无限大；景区周边村镇的居民以劳动力参股的形式参与了景区的旅游开发，黄山市尤其是市政府所在地屯溪区一直是黄山对外开放的门户和旅游者的集散地，对旅游发展基本持支持和乐观的态度，对游客的接纳程度较高，目前乃至今后一段时间本地居民的心理承载力将不会构成黄山风景名胜区旅游发展的瓶颈。

（2）游客心理容量

游客基于人群敏感的心理承载力因游客性格和行为特征的差异而有所不同。游客的心理阈值往往产生于景观观赏要求高、知名度高的景区景点，而对于一般景区或者引景地带较为宽容，测算时只考虑游客对黄山内核心景区的心理承受容量进行计算，取游客对旅游空间的心理极值为 1.8 $m^2/$人，对北海、天海、西海、玉屏四个

景区的容量进行测算，心理容量极值为 135 976 人。

3.2.5 旅游环境容量

由上述可推测，夏半年天气条件正常的情况下，黄山风景名胜区的旅游环境容量如下：

$$TEBC_{夏} = \min\{64\ 846,\ 7\times10^4,\ 117\ 800,\ 135\ 976\} = 64\ 846\ 人/d$$

可见黄山风景名胜区的空间环境容量为景区旅游容量的卡口值。冬半年由于各景区的旅游时长缩短（10 h），游客周转率发生变化（游览速度减慢），大致估算冬半年每日旅游环境容量为：$TEBC_{冬} = 48\ 660$ 人/d。

3.3 黄山风景名胜区旅游环境承载与评价

3.3.1 日承载情况分析

（1）非游览高峰日的合理游客容量

以《风景名胜区总体规划标准》（GB/T 50298—2018）中规定的合理人均占有面积标准，按平均游览日周转率来估算各景区非游览高峰日的游客容量，得出结果如表 3-7 所示，景区的平均非游览高峰日合理游客容量约为 34 807 人。

表 3-7 黄山风景名胜区各景区的非高峰日合理游客容量估算　　单位：人次/d

景区	温泉景区	云谷景区	玉屏景区	北海景区	天海景区	西海景区	钓桥景区	东海景区	合计 2
步行道容量	271	1 464	2 694	1 815	4 754	1 669	8 332	5 784	26 783
机动车线路容量	691	—	—	—	—	—	1 156	291	2 138
平台景点容量	—	302	509	2 058	971	785	1 011	250	5 886
合计 1	962	1 766	3 203	3 873	5 725	2 454	10 499	6 325	34 807

（2）游客容量承载力与调控

根据《风景名胜区管理通用标准》（GB/T 34335—2017）规定，以合理游客容量和极限游客容量为基础进行游客量的调控。黄山的游览高峰日和非游览高峰日的极限游客容量分别为

$$TEBC_{夏} = 64\ 846\ 人/d \quad TEBC_{冬} = 48\ 660\ 人/d \quad TEBC_{平均} = 34\ 807\ 人/d$$

如表 3-8 所示，冬半年旅游高峰日游客人数达到 3.4 万，夏半年高峰日游客人数

达到 4.5 万，非游览高峰日游客人数达到 2.4 万时，景区应在智慧平台上发布预警公告，当人数达到红色预警范围时，应采取游客分流或分批进入景区等应急响应措施，保证游客安全。

表 3-8　黄山风景名胜区游客安全预警等级与人数

预警类型	预警等级	预警指标	预警人数/人
冬半年游览高峰日极限游客容量	黄色预警	≥$TEBC_{冬}$的70%	34 062
	橙色预警	≥$TEBC_{冬}$的80%	38 928
	红色预警	≥$TEBC_{冬}$的90%	43 794
夏半年游览高峰日极限游客容量	黄色预警	≥$TEBC_{夏}$的70%	45 392
	橙色预警	≥$TEBC_{夏}$的80%	51 877
	红色预警	≥$TEBC_{夏}$的90%	58 361
非游览高峰日合理游客容量	黄色预警	≥$TEBC_{平均}$的70%	24 365
	橙色预警	≥$TEBC_{平均}$的80%	27 846
	红色预警	≥$TEBC_{平均}$的90%	31 326

3.3.2　年承载情况分析

以此值估算黄山风景名胜区全年接待游客的总容量=34 807 人/d×365 d=12 704 555 人。按照《风景名胜区总体规划标准》（GB/T 50298—2018），年承载指数的计算公式如下：

$$年承载指数 = \frac{年接待游客人数}{年旅游环境容量}$$

2017 年黄山风景名胜区接待游客人数有 3 368 688 人，年承载指数约为 0.3，按目前评价指标为弱载程度。

3.3.3　承载压力评价

（1）空间容量超载带来的生态承载压力不容忽视

山上各个景点游客到达率不同，造成一些景点到达率过高，带来环境压力而成为全山容量的瓶颈：例如，北海景区现在基本上游览率为 100%，高峰时会出现严重超载现象，北海的旅游环境容量极易成为整个景区的环境容量卡口；玉屏、天海景区由于有索道、交通便利、地理位置优势等条件，属于游客集散地，高峰期也处于超载状态。

旅游高峰期北海、玉屏、天海等空间容量超载的景区，同时也会面临比较大的生态承载压力，大量游客拥堵、滞留，有可能造成山路沿线的植被踩踏、旅游垃圾瞬时排放量过高、森林火险等级上升等，对黄山的自然生态系统造成潜在的破坏。

（2）各类承载能力之间此消彼长

景区的基础硬件设施逐渐完善，玉屏、太平、云谷等索道的陆续开通，西海观光缆车通行，温泉公路班车运送等缩短了游客观光所需的时间，加快了游客群体的周转率，同时不断改变着客流方向，使得客流集中的时间、地点和程度有所变化，交通承载能力不断提升；景区内供水、供电系统不断完善，供给能力提高，设施环境容量也在不断扩大。

与此同时，游客人数激增带来极大的环境污染压力，固体废物垃圾产生量巨大，靠人力上下山背送，排污处理能力非常有限；住宿供水设施的扩建极易引起植被覆盖率降低和土壤质地的改变；施工期间的垃圾、粉尘和污水加剧了景区的生态环境压力；索道建设与运营期间，会对自然植被、树木生长和后期的植物发育造成影响，改变地下水循环规律，对动物生活习性也造成一定干扰，景区的自然生态环境容量会随之降低。

（3）各景区的旅游环境承载容量不均

黄山风景名胜区旅游环境承载力全年综合利用水平为弱载，旅游淡季严重弱载，旅游旺季基本适载或轻微超载，这表明黄山的超载问题是时间和空间分布不均的问题，解决的措施不应只是如何扩大高峰期的承载能力，而是如何利用其余时期和其他景区的剩余容量，以及在高峰期如何分散超载景区的游客压力。

目前黄山风景名胜区的八个景区中，北海、西海、玉屏、天海景区的综合承载力利用状况超载，高峰期极易出现超载现象，成为游客容量卡口；温泉、云谷景区不容易出现超载，存在分量承载力的不协调现象，但客流周转频繁，生态承载力不高，可以考虑生态修复工程建设；西海大峡谷与松谷景区也存在分流游客不协调的现象，客流停留时长不够，大多数游客通过索道和观光缆车结束行程。

4 黄山风景名胜区生物多样性保护空间分区

4.1 黄山风景名胜区生物多样性保护目标分析

黄山风景名胜区是以山岳景观、古树名木为主体的风景名胜区，结合该区域黄山—怀玉山生物多样性保护优先区域的现状，黄山风景名胜区生物多样性保护的目标应包括：①珍稀濒危野生动植物及其生境；②自然灾害敏感区；③以古树名木为主的自然人文景观。

4.1.1 物种生境

在黄山风景名胜区动植物多样性资源分布特征的基础上，确定以云豹、黑麂为动物多样性保护的关键种，兰科植物为代表的珍稀濒危植物保护关键种。依据生态位理论，筛选海拔、生态系统类型等指标，识别目标物种的分布范围，为精确保护野生动植物栖息地提供依据。

4.1.2 自然灾害敏感区

自然灾害频发的地区一般是生态环境脆弱或受人类活动影响较大的区域，在黄山风景名胜区范围内识别灾害敏感区域，对有害生物防治、森林火灾等防控具有积极的意义。同时，对于景区内灾害的及时发现、抑制扩散和采取必要措施等也起到了推动作用。

4.1.3 自然人文景观

人类活动改变生物多样性的生存状态是由文化价值观决定的，文化表达人与生物多样性的互动关系，文化也是生物多样性保护的可靠社会力量。黄山生长的古树名木都是在历史的沉淀中遗留下来的宝贵财富，它象征着黄山历史文化的精髓，也是历代文人墨客所吟咏歌颂的对象。因此，保护黄山的古树名木也是对黄山历史文化与人文精髓的保护。由于黄山的古树名木有不少分布于旅游风景区，既要弘扬其文化，与游客接触，又要保护其不被游客破坏，精神传承对黄山显得尤为重要。

4.2 黄山风景名胜区生物多样性保护重要区域划定

4.2.1 划定方法

划定黄山风景名胜区生物多样性保护重要区域，可以保护生物生态系统的完整性、特殊物种栖息地和重要遗传资源分布地。重要区域的识别，需要对黄山生物多样性保护进行两个层次的评价与识别，包括：①生物多样性保护功能重要性评价与识别，获得生物多样性保护极重要区域；②保护物种栖息空间评价与识别，获得保护物种高适应性栖息地区域。通过评价结果，扣除重叠区域，联合极重要区域和高适应性区域获得黄山风景名胜区的生物多样性保护重要区域，见图4-1。

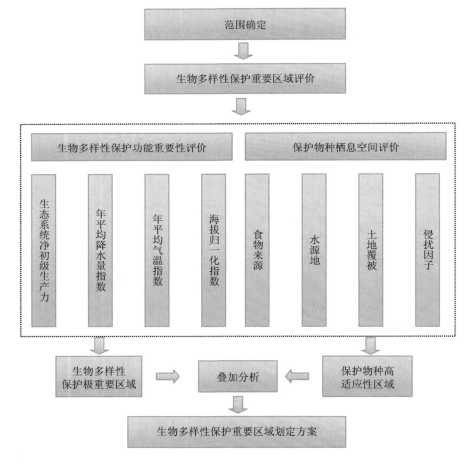

图4-1 黄山生物多样性保护重要区域划定技术路线

（1）生物多样性保护功能重要性评价

2015 年 4 月 30 日，环境保护部印发《生态保护红线划定技术指南》（环发〔2015〕56 号），适用范围为中华人民共和国境内生态保护红线的划定。黄山风景名胜区生物多样性保护重要区域的划定参照生态保护红线的划定方法，生物多样性保护功能重要性评价方法见表 4-1。

表 4-1 黄山生物多样性保护功能重要性评价指标

评价项目	评价模型	计算说明	
生物多样性保护功能重要性评价	基于生境多样性的方法 $S_{bio}=NPP_{mean} \times F_{pre} \times F_{tem} \times (1-F_{alt})$ S_{bio} 为生物多样性保护服务能力指数	NPP_{mean} 评价区域多年生态系统净初级生产力平均值	多年 MODIS 250 m 每 16 天合成 NDVI 数据产品
			多年 Landsat 或资源 3 号卫星等高分辨率卫星数据产品
			植被类型分布图
			气象站多年逐月太阳总辐射量
			气象站多年逐月太阳净辐射量
			气象站多年逐月降水总量
			气象站多年逐月蒸发总量
			气象站多年逐月平均地面温度
			气象站多年逐月平均气温
			气象站多年逐月平均最高气温
			气象站多年逐月平均最低气温
			气象站多年逐月平均气压
			气象站多年逐月平均水汽压
			气象站多年逐月平均风速
			气象站多年逐月平均相对湿度
		F_{pre} 为多年（大于 30 年）平均年降水量数据插值并归一化到 0～1	气象站多年逐年降水总量
		F_{tem} 为多年（10～30 年）平均气温数据插值归一化到 0～1	气象站多年逐年平均气温
		F_{alt} 为评价区海拔归一化到 0～1	DEM 数字高程模型

（2）物种栖息地识别

保护物种栖息空间评价是以明确界定目标物种生态位为原理，以保证物种的种群可持续发展为目的的评价方法。评价因子包括自然因素和人为因素，自然因素包括：食物源（空间插值）、土地覆被（LUCC）、海拔（DEM）、坡度、坡向、水源地；人为因素包括：居民点、道路及采矿区。该方法不仅是保护物种现有的栖息地，而且在一定范围内给物种群数量的增加预留了空间。

（3）景观视域区识别

黄山风景名胜区的最核心功能是旅游观光，保护景区游览区域的视域景观完整性是最基本要求，也是景区景观多样性的底线。以先规划的旅游路线为结点，通过GIS模拟游客视域范围，识别景观视域区。

（4）自然灾害敏感区

自然灾害是生物多样性锐减的重要因素，防范自然灾害是保护区域生物多样性的重要手段，识别黄山风景名胜区自然灾害敏感区将有效提高风险防范效率。依据黄山面临的主要自然灾害，采用GIS的手段计算出景区内灾害高敏感区域。

（5）生态廊道识别

从生物保护的角度出发，生态廊道为供野生动物使用的狭带状植被，以促进两地间生物的运动。建立生态廊道是景观生态规划的重要方法，是解决当前人类剧烈活动造成的景观破碎化以及随之而来的众多环境问题的重要措施。利用最短路径算法和GIS的空间分析功能，对黄山风景名胜区动物迁徙廊道进行识别与构建。

4.2.2 划定方案

（1）评价对象

在不影响景区主体功能的前提下，为实现不同层次的保护目标，体现：①保护生态系统完整；②预留物种栖息地；③景观不缺失；④预防自然灾害风险的管控要求。

依托现有遥感数据及地面监测数据，应用地理信息系统，识别出生物多样性保护屏障区、保护物种高适应性区域、景观视域区、自然灾害敏感区和生态廊道，并提出分类管控措施。

（2）评价过程

1）生物多样性保护功能重要性评价

利用地理信息系统软件，将生物多样性保护功能服务值在黄山风景名胜区层面采用quantile（分位数）方法进行4级分类（classified）操作。按生态系统服务值由低到高依次划分为一般重要、中等重要、重要、极重要4个级别，将重要区域提取生成生物多样性保护屏障区，见表4-2。

表 4-2　黄山生物多样性保护功能重要性评价方法

评价项目	评价模型	模型因子计算说明
生物多样性保护功能重要性	$S_{bio}=NPP_{mean} \times F_{pre} \times F_{tem} \times (1-F_{alt})$ S_{bio} 为生物多样性保护服务能力指数	NPP_{mean} 为评价区域多年生态系统净初级生产力平均值 F_{pre} 为多年（大于 30 年）平均年降水量数据插值并归一化到 0~1 F_{tem} 为多年（10~30 年）平均气温数据插值并归一化到 0~1 F_{alt} 为评价区海拔归一化到 0~1

根据评价结果，黄山风景名胜区内生物多样性保护重要区域分布在景区的绝大部分区域，面积约为 134.78 km²，约占景区总面积的 83.92%；与黄山市生物多样性保护重要区域对比可以看出，黄山风景名胜区是黄山市西北部重要的生物多样性保护功能区。

2）保护物种适应性分析

根据生态位模型，综合应用 GIS 技术和层次分析法，选定黄山 I 级保护动物黑麂、云豹（专栏 4-1，专栏 4-2）、兰科植物为黄山风景名胜区生物多样性保护重要区的研究对象，厘定其高适应性区域。

专栏 4-1　黑麂

黑麂（*Muntiacus crinifrons*），别称乌金鹿、蓬头鹿属，是麂类中体型较大的种类。国家 I 级保护野生动物，《濒危野生动植物种国际贸易公约》（CITES）附录 I 物种，世界自然保护联盟（IUCN）将其濒危等级列为易危。黑麂体长 100~110 cm，肩高 60 cm 左右，体重 21~26 kg，冬毛上体暗褐色，夏毛棕色成分增加；尾较长，一般超过 20 cm，背面黑色，尾腹及尾侧毛色纯白，白尾十分醒目[①]。

① 黑麂图片来源：中科院植物所古田山国家级自然保护区红外拍摄资料，http://www.jiaodong.net/news/system/2009/06/16/010555671.shtml。

黑麂是中国的特产动物，没有亚种分化，分布范围十分狭小，分布于中国安徽南部、浙江西部、江西东部的怀远和福建北部的武夷山地区；栖息于海拔 1 000 m 左右的山地常绿阔叶林及常绿、落叶阔叶混交林和灌木丛。

　　黑麂胆小怯懦，恐惧感强，大多在早晨和黄昏活动，白天常在大树根下或在石洞中休息，稍有响动立刻跑入灌木丛中隐藏起来，其在陡峭的地方活动时有较为固定的路线，常踩踏出 16~20 cm 宽的小道，但在平缓处则没有固定的路线。

专栏 4-2　云豹

　　云豹（*Neofelis nebulosa*）为哺乳纲的猫科动物，有四个亚种，体长 70~110 cm，尾长 70~90 cm，体重 16~40 kg，为豹亚科最小者。云豹分布于亚洲的东南部，从最西部的尼泊尔开始，一直向东到中国台湾，包括缅甸和中国秦岭以南；往南则从印度东部、中南半岛开始，一直向南到马来半岛等地为止。云豹是高度树栖性的物种，经常在树木上休息和狩猎，但是它们在地面上的狩猎时间要比树上的更长。云豹栖息于亚热带和热带山地及丘陵常绿林中，最常出现在常绿的热带原始森林，但也能在其他的生境中看到它，如次生林、红树林、沼泽、草地、灌木丛和沿海阔叶林，垂直高度可达海拔 1 600~3 000 m，适宜环境温度在 18~50 ℃[①]。

　　依据分析目标物种的生活习性，栖息地分布的生态影响因子有 8 个，包括自然因素和人为因素，自然因素包括：食物源、土地覆被、海拔、坡度、坡向、水源地；人为因素包括：居民点、道路。通过对以往文献的调研和对保护专家的问询，可得知保护物种的最佳海拔高度，距水源地最佳距离，对不同土地覆被类型的选择，栖息坡度、坡向，距人为干扰因素的距离。

　　根据 8 个生态因子（变量）对不同物种的影响情况，分别赋予不同的数值（取值范围为 0~100），每个生态因子中分值为 100 的区域为最佳栖息地。

　　利用 GIS 空间分析功能，将各因子根据权重值进行叠加运算，得到保护物种最

---

① 云豹图片来源：http://tupian.baike.com/ipad/a1_55_75_01300000174719121437758683327_jpg.html。

佳栖息地分布图，将计算结果分类，提取保护物种高适应性区域。根据分析结果，黑麂的高适应性区域分散分布，生境受道路影响较明显，极度适应区面积约占景区总面积的 33.62%，主要分布在景区东部及西部，这些区域有大面积完整生境，其中浮溪管理区中部、钓桥管理区南部、松谷管理区西部和福固管理区西部是开展黑麂监测和就地保护关键区域。该区域森林生态系统完整，水源充足，离游客距离较远，人为干扰较小，有利于黑麂种群的恢复（图4-2）。

根据分析结果，云豹的极度适应区集中分布在中部及西部区域，极度适应区面积约占景区总面积的 8.92%，该区域森林生态系统完整，水源充足，离游客距离较远，人为干扰较小，有利于云豹种群的回归与恢复。根据分析，云豹生境受道路影响较明显，其中浮溪管理区中部、钓桥管理区东部、玉屏管理区等区域是云豹的极度适应区，可能活动的概率较高，是开展云豹就地保护与监测的重要区域（附图4）。

根据分析，黄山风景名胜区森林覆盖率高，原生植被保存完整，以兰科植物为代表的珍稀濒危植物保护工作成效较好，极度适应区域范围较大，面积约占景区总面积的 31.91%，主要分布在温泉管理区南部、洋湖管理区北部、福固管理区东部及浮溪管理区西南部等。另外，黄山风景名胜区实行"一树一策"的保护措施，旅游道路附近的名木古树得到有效保护，古树资源保存完整（图4-2）。

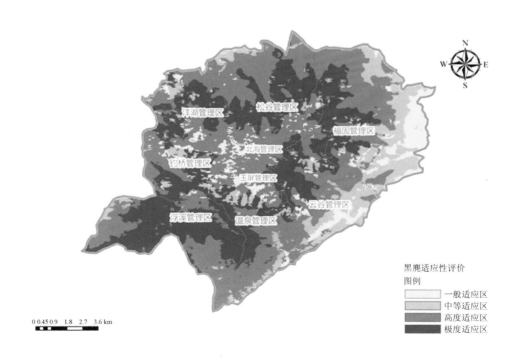

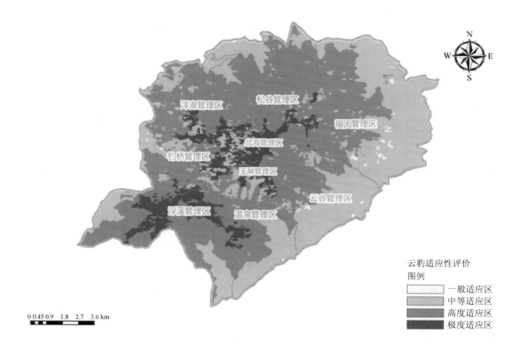

图 4-2　黄山风景名胜区重要物种生境与古树分布适应性评价

（3）景观视域分析

基于黄山风景名胜区 DEM、游览线路图和其他景区规划数据，模拟游客步行可观赏的景观范围，识别景区的可视范围，形成景观视域区，总面积约为 85.43 km²，约占景区总面积的 53.19%。识别区域主要分布于北海管理区、松谷管理区、玉屏管理区、云谷管理区和温泉管理区，涵盖了景区的主要景点范围和野外探险区及其他景点等。

（4）自然灾害敏感性分析

黄山风景名胜区为地形强烈切割区，切割深度多在 500～1 000 m，悬崖峭壁，凸面山坡，坡陡峰险，险峻处坡度在 60°～74°，土层稀薄，容易成为水土流失的敏感区。分析结果显示，黄山风景名胜区水土流失极敏感区约占总面积的 8.67%，主要分布在玉屏管理区、北海管理区和云谷管理区等区域。

黄山主要植被类型是针叶林，同时针叶林也是黄山重要的景观主体，容易受到松毛虫和松材线虫的侵害。依据预防优先的原则，根据现有数据，通过空间运算，识别出松材线虫灾害极敏感区约占景区总面积的 14.52%，主要分布在云谷、福固管理区的低海拔区域。

森林火灾是景区的重点防范灾害。根据景区提供的数据，识别出一级、二级、三级火灾敏感区域。一级区域主要分布在核心景区周边（图 4-3）。

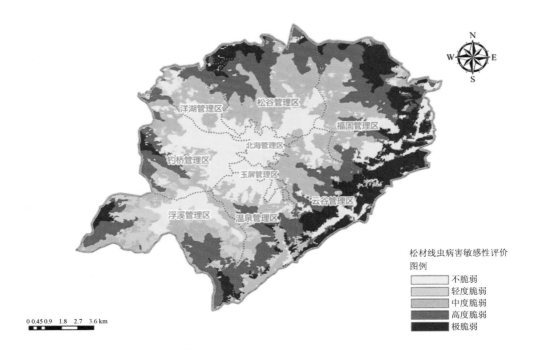

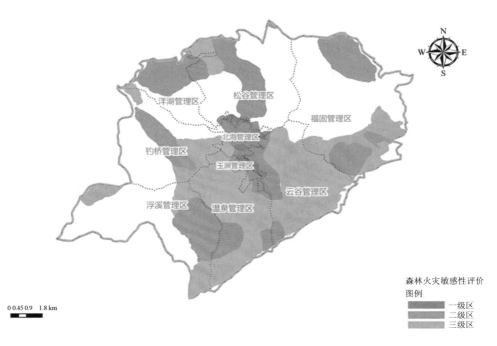

图 4-3 黄山风景名胜区自然灾害敏感性分析

4.2.3 重要区域方案集成

通过评价结果，扣除重叠区域，与现有《黄山风景名胜区旅游发展总体规划》等规划相衔接，联合极重要区和高适应区域，得到黄山风景名胜区生物多样性保护重要区域（图4-4），见表4-3。

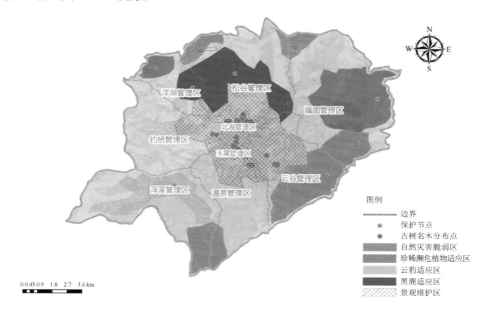

图4-4　黄山风景名胜区生物多样性保护重要区域分布示意

表4-3　黄山风景名胜区生物多样性保护重要区域名录

	重要区域类别		分布范围	面积/km²
I	生物多样性保护屏障区		景区大部分范围	134.78
II	保护物种高适应区	黑麂适应区	松谷、福固、云谷和浮溪管理区	60.80
		云豹适应区	浮溪和钓桥管理区	35.20
		濒危植物适应区	福固和云谷管理区	56.00
III	景观视域区		松谷、北海、玉屏、云谷和温泉管理区	85.43
IV	自然灾害脆弱区		北海、玉屏、福固和云谷管理区	91.78

4.3　黄山风景名胜区生物多样性保护分区管控措施

4.3.1　生物多样性保护屏障区

该区分布在景区大部分自然生态系统区域，面积较大。该区域需要保证生态系统的完整性，生态服务不下降，并开展定期评估，减少人为活动干扰。

4.3.2 保护物种高适应区

保护物种高适应区是珍稀濒危物种保护与恢复的珍贵生境，黑麂的高适应区主要分布在松谷、福固、云谷和浮溪管理区；云豹的高适应区分布在浮溪和钓桥管理区，濒危植物高适应区主要分布在福固和云谷管理区。

该区域远离人为活动区域，严格禁止游客进入，周边避免人为建设活动，以免惊扰动物。在高适应区内应开展物种调查与目标物种监测，定期评估目标物种（云豹、黑麂等）的种群状况。可根据自然规律、不同季节、昼夜变化等开放与封闭该区域，以保护高适应区的生物资源。

4.3.3 景观视域区

景观视域区是游客游览黄山的主要景观区域，直接影响游客的观感效果，该区域主要分布在松谷、北海、玉屏、云谷和温泉管理区。该区域要注重保护自然景观的完整性，尽量避免修建不协调的人造景观。该区域在后期规划中，需要注意建设与景观的协调关系。

4.3.4 自然灾害敏感区

黄山风景名胜区识别出的自然灾害敏感区主要面临水土流失、雷击、林火和林木虫害的风险，该区域主要分布在北海、玉屏、福固和云谷管理区。针对山麓地区容易遭受雷击引发的林火灾害，做好火灾预警与防护；定期监测景区主要病虫害情况，防护优先。

4.3.5 时间管控

根据《黄山风景名胜区景点封闭轮休规范 基本要求》（DB 34/T 1241—2010），对黄山已经开发旅游风景区的区域采取轮休开放。通过人工措施辅助自然恢复的方式，恢复林下植被，促进树木生长，改善生态环境，使景区实现可持续发展。着重考虑动物因季节交替的迁徙活动，根据周期性的迁徙活动调整时间管控措施，达到协调发展与保护的目的。

5 黄山风景名胜区生物多样性保护面临的问题

5.1 环境承载力分布不均，自然生态面临瞬时压力

采用旅游环境容量静态模型（图5-1），结合2018—2030年景区发展建设规划，从空间环境、生态环境、经济环境和社会心理环境容量等对景区的旅游环境容量进行测算和分析（附表6、附表7），得出以下主要结论。

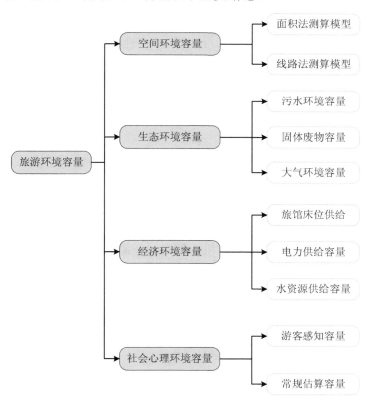

图 5-1 黄山风景名胜区旅游环境容量静态模型示意

43

5.1.1 高峰期核心景区极易出现容量超载

目前，黄山风景名胜区全年旅游环境承载力综合利用水平为弱载，但各景区的旅游环境容量各不相同。北海、西海、玉屏、天海景区的旅游环境容量总体低于东海、钓桥等景区，综合承载力利用状况极易超载；北海景区基本游览率为100%，高峰时会出现游客滞留现象，极易成为整个景区的环境容量卡口；核心景区由于索道、交通、线路密集成为游客集散地，高峰期极易出现游客拥堵滞留。

5.1.2 瞬时游客规模对景区自然生态保护造成压力

高峰时段黄山核心景区的瞬时游客量激增，人群缺乏有效导引和分流，极易造成景区交通线路和旅游步道拥堵，有可能造成山路沿线的植被踩踏和破坏、旅游垃圾瞬时排放量过高、森林火险等级上升，对黄山的自然生态系统造成潜在威胁，基础设施、索道缆车达到运载上限，固体废物和污水排量巨大，供水供电量超负荷，进一步加剧自然生境的受干扰程度，黄山自然生态系统和景观完整性以及生物多样性的生态保护价值将受到巨大影响。

黄山风景名胜区面临的超载问题主要体现在环境容量的分布不均，需要重点解决游览高峰期如何分散超载景区的游客压力，保障自然生态系统免受过多干扰的同时向游客提供优质的生态旅游服务。

5.2 生物多样性管护任务艰巨，监测预警能力薄弱

5.2.1 生物多样性本底数据不清

景区生物多样性本底数据尚未形成体系，云豹、金猫、大灵猫、豹、狗獾等国家和安徽省级重点保护动物基础数据匮乏，保护工作难度较大；收集景区内食物链顶端物种的数量、分布区域、栖息环境等信息迫在眉睫；白颈长尾雉等珍稀鸟类观测难度大；大鲵、棘胸蛙、平胸龟、尖吻蝮等两栖爬行类动物本底不清；昆虫、微生物类资源尚无基础数据。除此之外，景区内的普通关键种、特有农作物等特色生物资源亟待系统调查与研究。

5.2.2 松材线虫等病虫害及外来物种入侵形势严峻

松材线虫病目前正威胁着黄山松的生存，黄山风景名胜区具有适宜该病害发生的各项条件，一旦松材线虫病在景区暴发，将可能对该地区的人文、自然景观造成严重破坏，随时都有可能对景区的生态资源和人类宝贵的自然遗产造成不可挽回的

巨大损失。

已有调查研究发现，景区约有 31 种外来入侵物种分布范围较广并造成一定危害，加拿大一枝黄花、美国白蛾等外来物种已呈现明显的种群暴发趋势。

5.2.3 监测与灾害预警能力薄弱

景区生态环境脆弱，自然灾害频发且破坏力强，森林火灾极易发生，资源本底数据缺乏有效整合，需要进一步对景区内珍稀濒危物种、种群数量、分布地点、生长状况等内容进行系统整合，同时考察社会、经济等因素的影响下景区生态环境的变化趋势。

5.2.4 专业管护人员及技术手段相对匮乏

景区在生物多样性保护与管理方面面临的最大的制约因素就是缺少相关的技术人才，人员对专业设备和仪器的操作能力不足，设备在操作和长期运行维护上面临困难，未形成有效合力；对国内外游客的生态环境保护宣传教育方式相对单一和粗浅，缺乏先进的生态环境保护宣教场馆和宣教设施，宣教工作经验尚显不足，需要开展有针对性的专业技能培训，积极参与国际交流，汲取国际先进经验。

5.3 资源可持续利用尚未实现，社区集聚发展效率不高

黄山风景名胜区生物多样性的经济价值总量极高，分为直接开发价值、生态保护价值和未来存在价值三部分，主要存在以下问题与挑战。

5.3.1 门票和餐饮等直接收入占总收入比重较高

门票和餐饮住宿收入占目前黄山风景名胜区生物多样性直接开发价值的绝对比重，黄山生物多样性所蕴含的深厚历史文化价值没有得到充分的物化表现，产品价值的附加值较低。景区生物多样性的直接开发价值如果长期单纯依赖门票收入，增加收入的方式只能依赖增加游客人数，而人数的不断增加对景区生态环境和游客的旅游体验会造成巨大压力，不利于景区的可持续发展。

5.3.2 景区发展对周边社区的带动能力尚显不足

当前景区旅游以观光游览为主，功能相对单一，对相关附属产业的辐射与带动功能较弱，在特产销售、乡土文化、人文游历等方面，游客参与和体验的机会不足，回头客较少，重游率较低；缺乏生物资源开发利用的统筹和系统规划，景区周边社区居民的生产经营活动大多为自制自销、零星售卖，利润率低，受市场风险影响大，

发展效率不高，此外分散式经营还容易造成生物资源的无序开发和浪费，不利于景区的生态环境及动植物资源保护。

5.4 地域性传统文化缺乏传承，品牌价值效应尚未显现

黄山生物资源价值蕴藏量巨大，具备独特的地理和文化附加值，当地特色资源、农产品和相关文化产品种类繁杂，品质不一，缺乏统一和正规的包装和商标，尚未形成产业规模，亟须增强黄山（徽州）产业品牌集聚效应；对徽州地域特色文化的关注和挖掘利用力度不足，传统文化，尤其是以生物资源为原料、科学利用生物资源的传统工艺传承能力不断削弱，是需要引起高度重视的问题之一。

黄山生物多样性所蕴藏的巨大自然人文遗存价值需要引发公众强烈的文化认同感，亟待创新黄山（徽州）文化的传播传承方式，借助文化载体激发社会公众对黄山自然生态与生物资源保护的使命感。

6 黄山风景名胜区生物多样性保护行动计划

6.1 保护目标与主要任务

6.1.1 总体目标与阶段性目标

按照建设"生态最佳、环境最美、保护最好、管理最优、品质最高、品牌最靓"的"美丽黄山"的要求，严守生物多样性保护底线，以生物多样性保护优先行动为抓手，确保景区生态系统服务不降低、受保护空间面积不减少，提升区域生态系统服务功能的同时促进黄山景区可持续发展，促进生物多样性保护与景区经济社会发展的协调统一，最终实现人与自然的和谐共处。

（1）近期目标

到2022年，生物多样性保护能力得到大幅提升。完成全景区系统的生物多样性本底调查与评估；基本建立生物多样性监测和预警网络体系，所有站点能开展生物多样性的日常监测；加强就地保护和迁地保护，建设一批保护小区和生物物种资源种质库、圃，景区管护能力得到较大程度提升；生物多样性保护相关法规、制度得到完善，争取成为国家公园试点单位。

（2）中期目标

到2025年，生物多样性保护能力得到显著增强。各类生态系统、全部国家和省级重点保护与珍稀濒危野生动植物得到有效保护；生物多样性的可持续利用能力不断加强，实现生物多样性保护利益相关方的惠益共享；国内与国际合作广泛开展，国际知名度和影响力大幅提高；生物多样性保护宣传教育和科技得到普及，公众积极参与保护生物多样性。

（3）远期目标

到2030年，生物多样性得到切实保护。各项管理制度和保护政策成效明显，生物多样性管护能力和科研水平极大提高，生态系统处于良好状态，全景区物种资源得到有效保护，遗传资源得到有效保存，能满足景区可持续发展的需要，公众自觉参与生物多样性保护，实现人与自然和谐相处。

6.1.2 生物多样性保护的主要任务

（1）开展空间分区，构建重要区域分布格局

建立黄山生物多样性保护重要区域划定方案。依照《生态保护红线划定技术指南》（环发〔2015〕56号），对景区的生物多样性保护重要功能、保护物种栖息空间、景观及自然灾害敏感地区进行识别和评价，力求生态系统保护完整，重要物种栖息地得以预留，景观不缺失，自然灾害风险得到有效预防。

构建重要区域的空间分布格局。以《黄山风景名胜区总体规划（2007—2025年）》为依据，划定黄山风景名胜区生物多样性保护的重要区域，构建景区生物多样性保护空间格局，重点保护景区生态系统的完整性、旅游生态环境的安全性、特殊物种栖息地和重要遗传资源分布地。

（2）确定保护目标，落实分级分类管控措施

设置保护目标，实施分级管控。以"红线管控，预留空间"为指导思想，对景区生物多样性保护空间格局进行分级管控。以重要物种及其生境、生态脆弱区、自然灾害敏感区等保护关键节点为一级红线区，以生物多样性保护屏障区为二级核心区，以景观视域带为三级优化开发区。

确定承载红线，开展分类管护。对重要物种及其生境、生态脆弱区、自然灾害敏感区等保护关键节点设置管控红线，严格限制资源开发和人类活动行为，确保受保护面积不减少；对生物多样性保护屏障区内受损生态系统进行自然或人工修复，确保生态功能不降低；对四条景观视域优化带进行优化开发，预留未来发展空间；设置游客承载和环境消耗红线，确保资源使用不超线。

（3）科学利用资源，联合周边社区共同发展

推进生物资源保护与利用。以"物尽其用，着眼未来"为原则，推动黄山风景名胜区的生物资源以及与生物多样性相关的人文资源的保护与可持续利用，对黄山松、黄山杜鹃、黄山短尾猴等珍稀濒危特有物种进行电子标识，完善迁地繁育基地和遗传资源库建设；对黄山茶、菌类、笋类、药材等景区重要且具有开发前景的生物资源进行规范化管理和原始种保护，与地区特色文化资源进行有机结合，建立科学的市场开发机制；对景区菌类等采取禁采等措施，降低资源过度开发程度；积极开展外来入侵物种防治和技术研究。

联合周边社区参与协同发展。景区协调周边"五场一镇"共同打造黄山风景名胜区特色旅游品牌，合力推进黄山"大旅游"产业建设。以彰显徽州文化和"世界自然人文双遗产"的地理标识和地域特色为手段，建立黄山旅游生态商标和服务标

志，延伸旅游产业链条，融入徽派建筑、民俗工艺、特色物产、徽派饮食等传统文化知识，实现山上山下"统一商标、统一制服、统一服务、协调共管"，凸显黄山旅游品牌的优势效益，实现景区和周边社区协同共赢发展。

（4）推广旅游宣传，打造景区特色生态品牌

加强景区生物多样性保护宣传教育。采用多渠道、多手段推动公众宣传，提高公众保护意识，建立信息公开和公众监督、举报机制，完善公众参与保护的途径；要求并支持协助各旅行社加强游客宣教，建立健全景区多语种导览系统，向国际国内公众展示景区自然及人文历史、动植物资源和文化艺术作品等；依托景区商店、微信、微博、互联网电商等渠道，大力推广黄山文创产品及生态商品。

深化生物多样性保护合作伙伴关系。发展、深化国内与国际合作与交流，吸引国际、国内相关部门、科研机构、非政府组织、企事业单位、社区、社会公众等力量参与生物多样性保护项目的组织与实施；高度重视大中专院校、中小学的教学科研、自然教育等基地建设，以发展眼光培养长期伙伴关系，树立黄山风景名胜区生态旅游品牌。

（5）制订专项计划，拓宽资源保护利用途径

制订黄山生物多样性保护优先行动计划。以《中国生物多样性保护战略与行动计划（2011—2030年）》《黄山风景名胜区总体规划（2007—2025年）》为蓝本，结合黄山生物多样性保护的需要，针对不同保护目标和保护重要性，制订黄山生物多样性保护优先行动计划，明确各项保护计划，分批次、分主次加以实施。

采取多种途径推进生物多样性保护。加强退化生态系统恢复、珍稀濒危物种的种群扩增、就地与迁地保护以及物种资源的科学开发利用；构建动态监测与预警平台，推进物种资源信息化管理；强调景区与周边社区协同发展，注重专业技术人才培养与培训机制建设，实施全方位、多角度的生物多样性保护与管理策略，保障黄山区域生态安全和旅游产业的可持续发展。

6.2 黄山风景名胜区生物多样性保护优先行动

根据战略和阶段目标，确定景区生物多样性保护的10个优先领域及30个优先行动。

优先领域一：开展生物多样性调查、监测与信息化管理

行动1——生物多样性本底调查、编目与评估

①系统开展景区的重点物种资源本底综合调查，进行资源档案建立和编目。②采用"3S"技术，结合实地调查，在景区全域范围内开展植被状况调查，编制植被分布图。③对景区重要生态系统和生物类群的分布格局、变化趋势、保护现状及存在

的问题进行评估，发布综合评估报告。④开展景区生态系统服务、物种资源经济价值的评估研究，探索建立价值评估体系。

行动2——特色生物物种资源的调查编目

①开展景区及周边社区畜禽、水产品种资源的调查和收集整理，进行资源档案建立和编目。②开展景区重要林木、野生花卉、药用生物、野生蔬菜、农业及近缘植物等种质资源的专项调查，查清物种资源的种类、数量、分布和生境状况，进行资源收集保存、编目和数据库建设。

行动3——生物多样性保护相关传统知识的调查与编目

①开展对景区周边社区生物多样性保护相关传统知识的调查、记载、整理、编目，其习俗、信仰等可作为保护生物多样性和可持续发展的重要管理资源。②重点调查与整理当地社区对生物资源利用的传统地理特征产品、与生物多样性相关的传统技术和习惯用法等。

行动4——生物多样性监测与预警

①开展国家、省级重点保护物种的监测与预警。针对主要包括分布于景区的国家Ⅰ、Ⅱ级保护动植物及安徽省保护动植物，开展种群及其栖息地或生境的监测及预警。②开展重点区域黄山特有珍稀濒危植物及其生境的监测与预警。进行生境、群落动态等方面的监测与预警，研究受威胁的内在机制和外在因素，及时发现新的处于受威胁状态的物种。③开展重要生态系统的监测与预警。在保护重点区域选择具有代表性的森林生态系统开展监测与预警，及时掌握变化情况，科学地进行保护和管理。④开展旅游项目建设对生物多样性影响的监测研究。对景区的旅游活动进行长期监测研究，分析旅游开发对景区重要生态系统和重点保护动植物种的干扰影响。⑤完善基础设施，购置先进设备，逐步建立生物多样性数字化监测预警网络体系，实现数据共享。

行动5——生物物种资源信息化管理

建立黄山风景名胜区生物多样性信息管理系统。收集、整合现有生物资源信息，补充重点调查内容，建立包括物种分类、分布区域、资源储量、保护等级、受威胁和保护情况等为主要内容的物种资源数据库。

优先领域二：保护珍稀濒危野生物种资源

行动6——黄山短尾猴及其栖息地保护

①开展黄山短尾猴数量、分布区域和栖息环境的专项调查，确立黄山短尾猴的生态位，根据黄山短尾猴的生活习性分析制约其生长和繁衍的各项因素。②设立黄山短尾猴保护小区，保证野生状态的黄山短尾猴的种群数量。③开展黄山短尾猴的人工饲养和繁殖的研究，为野化放生做好前期工作。

行动7——云豹等食肉动物保护

重点调查云豹、金猫、大灵猫、黑熊、豺、狗獾等国家和安徽省级重点保护动物的数量、分布区域和栖息环境状况，根据它们的生活习性分析制约其生长和繁衍的各项因素；设立保护小区和生物廊道，建立常规保护制度并制定保护方案。

行动8——白颈长尾雉、白鹇等鸟类保护

重点调查白颈长尾雉、白鹇、勺鸡等国家和安徽省级重点保护鸟类数量、分布区域和栖息环境状况，设立保护小区和生物廊道，建立常规保护制度并制定保护方案。

行动9——大鲵、尖吻蝮等两栖、爬行动物保护

重点调查大鲵、棘胸蛙、平胸龟、尖吻蝮等国家和安徽省级重点保护动物的数量、分布区域和栖息环境状况，根据它们各自的生活习性分析制约其生长和繁衍的各项因素。设立保护小区，建立常规保护制度并制定保护方案。

行动10——黑麂等有蹄类动物保护

重点调查黑麂、獐、鬣羚等国家和安徽省级重点保护动物的数量、分布区和栖息环境状况，根据它们各自的生活习性分析制约其生长和繁衍的所有因素。设立保护小区和生物廊道，建立保护制度并制定保护方案。

行动11——黄山梅等珍稀濒危植物保护及种群扩增

对黄山梅、香果树、华东黄杉、金钱松、连香树、黄山木兰、天女花、黄山花楸等珍稀濒危植物进行重点保护，开展保护生物学研究，保护现有植株和种群，进行人工扩繁试验，逐步扩大人工种群。

行动12——扇脉杓兰等珍稀兰科植物保护及种群扩增

开展以扇脉杓兰、金兰、银兰等为代表的黄山兰科植物的野生资源专项调查，设立专门的野外保护点，进行人工繁育研究，开展野外回放试验，保护种质资源。

优先领域三：加强生态系统恢复

行动13——退化生态系统的恢复

对景区内受到人为活动严重干扰、服务能力退化且难以自然恢复的森林生态系统、河溪等淡水生态系统等进行恢复，制订恢复计划，按照最小风险与最大利益原则，通过工程措施和生物措施相结合，因地制宜地开展生物多样性恢复建设。

行动14——人工纯林生态系统的恢复

对树种单一、群落层次结构简单、生态功能低下的松、杉、竹类等人工纯林生态系统开展生物多样性的恢复，引入、培植乡土物种，通过人工措施促进自然恢复的方式，提升人工纯林生态系统的生物多样性和生态功能。

优先领域四：强化外来入侵物种防治

行动15——外来入侵物种防治

①加大对外来入侵物种测报点和检疫检查站的投入，加强外来入侵物种日常检疫监管，严格检疫申报和入境检查，防止人为传播。②制订外来入侵物种防治工作

方案，加强对松材线虫、松毛虫和松褐天牛等外来入侵生物的防治技术研究，对枯死松树集中进行无害化处理。

优先领域五：加强就地保护和迁地保护

行动16——提升景区就地保护的管理水平

健全机构，成立专门的生物多样性保护管理机构，学习国内外生物多样性保护管理方面的先进经验，借鉴国际知名景区在生物多样性管理体制、管理方式、资金投入机制等方面的具体举措，结合景区实际不断拓展深化，实现景区生物多样性保护管理水平的大幅提升。

行动17——强化古树名木保护

开展古树名木管护复壮技术研究，严格执行《黄山风景名胜区古树名木保护管理规范》《黄山风景名胜区古树名木复壮技术规范》，不断完善古树名木保护方案和灾害性天气应急防护预案，强化保护管理措施，建立综合保护体系，加大对建档保护外的古树保护力度。

行动18——科学合理开展迁地保护

开展景区珍稀濒危动植物种的繁育和恢复技术研究，着重开展黄山特有观赏花卉、药用植物等物种资源的人工培育和重要保护动物物种的人工养殖，减少对野生物种资源的破坏。建立1~2个迁地保护园、种质资源库（圃）或重要物种救护繁育基地。

行动19——环境综合治理

①继续实施景区环境的综合整治，完善垃圾、污水的集中处置设施。②推进景区周边村镇污水和垃圾治理，逐步开展周边社区污水、垃圾、农业面源污染、禽畜养殖污染的综合治理工作。

优先领域六：加快物种资源的科学利用

行动20——加强生物物种资源科学利用的创新研究

加快开展当地特色农作物、药用植物、观赏花卉、水生生物和畜禽遗传资源等生物资源开发利用技术和经营模式的创新研究，为景区生物资源的可持续利用提供科学依据。

优先领域七：促进当地社区可持续发展

行动21——替代生计示范与推广

选择景区周边农村社区建立示范试点，探索资金支持方式和资源开发方式，因地制宜地选择黄山茶、山野菜、药材、观赏花卉等特色资源进行市场化开发，在大型黄山主题活动设立社区特色商品销售窗口，大力发展生态旅游和电子商务，延伸旅游产业链条，带动社区居民就业、创业，推广示范经验，提高当地社区的整体发展能力。

行动22——替代能源示范与推广

开展替代能源调查及技术、经济可行性的研究，选择周边社区作为试点有针对性地进行替代能源的示范并进行推广，以减轻对植被的压力、对生态系统的干扰和对环境的污染。

优先领域八：完善政策法规与体制机制

行动23——建立和完善与生物多样性保护有关的政策法规

①完善与生物多样性保护相关的法律规章，制定和实施有利于生物多样性保护的相关政策和措施，特别是在投资、产业等方面的鼓励政策和措施。②制定和完善促进生物多样性保护的生态补偿机制，研究并制定资源开发与生态保护之间、景区与周边社区之间的生态补偿机制与标准，逐步建立生物多样性保护生态补偿标准体系。③制定并出台重大工程环境评价中必须执行生物多样性影响评价的制度规定，建立重大工程环评中生物多样性影响评价的公众参与制度。④制定并出台领导干部生物资源资产离任审计制度、生物多样性损害责任追究制度和损害赔偿制度。

行动24——创新生物多样性保护的体制与机制

①研究建立生物多样性保护协调机制，加强部门间的合作，促进景区各部门之间的信息交流和行动协作，保证生物多样性保护相关制度政策反馈渠道的畅通，建立健全打击破坏生物多样性违法行为的跨部门协作机制。②探索建立景区与当地社区的协作共管机制。

行动25——在部门规划、计划中强化生物多样性保护内容

①推动景区相关部门在制订部门发展规划、工作计划时加入生物多样性保护的内容，体现生物多样性保护要求。②建立生物多样性保护相关规划、计划实施的评估监督机制，促进各部门相关计划和规划的有效实施。

优先领域九：积极应对气候变化对生物多样性的影响

行动26——气候变化对生物多样性保护的影响评估与应对

①研究气候变化对景区重要生态系统、物种、遗传资源及相关传统知识的影响，重点研究气候变化对敏感区生物多样性的影响。②探讨应对气候变化的举措，制订并不断完善生物多样性保护应对气候变化的方案。③景区绿化选用抗逆性强的本土植物品种，增强植物适应气候变化的能力。逐步建立一定范围的结构合理、功能完善的生态系统序列，减轻气候变化对生物多样性的不利影响。

优先领域十：推进宣传教育与公众参与

行动27——加强生物多样性保护领域人才培养

①对景区相关部门管理人员和技术人员开展生物多样性知识和专业技术培训，提高其管理水平和专业技术能力。②建立科学有效的人才管理机制，吸引优秀科研人才加入景区从事生物多样性保护研究工作。③以丰富的生物多样性本底资料与较

高的生物多样性保护水平为基础，积极吸引人才与资金落地黄山，形成"人才—资金—技术"的良性互动。

行动28——加强生物多样性保护宣传和教育

①利用网络、报纸、电视、广播等各种传播媒体，对游客、社区群众开展生物多样性宣传和教育，大力宣传生物多样性保护的重要性，提高景区工作人员、游客、社区民众对生物多样性保护的认识，倡导有利于生物多样性保护的消费方式和餐饮文化。②加强生物多样性保护方面的法制普及工作，提高管理人员、游客和当地社区居民生物多样性保护的法律意识。③支持当地相关院校、环保社团、社会组织等开展生物多样性保护宣传教育，组织生态营地建设进行生态知识普及与教育、生态体验等。④举办优秀生物多样性艺术作品展，鼓励原创性生物多样性保护文化创意产品及影视、摄影、书画创作。

行动29——建立生物多样性保护公众参与机制

①研究公众在保护生物多样性方面的多方面需求，建立多渠道公众参与生物多样性保护的途径和机制。②建立多层次生物多样性保护合作伙伴关系，引导政府部门、企业、私营部门、公众、非政府组织以多种方式参与生物多样性保护。③提高公众参与水平，打造黄山动植物辨识云互动平台，利用大数据与人工智能技术，提高公众对生物多样性的保护兴趣与意识。

行动30——加强对外交流与合作

争取国际合作项目，引进综合生态系统管理、生态补偿机制、生物多样性保护廊道建设、气候变化与生物多样性保护等国际先进的管理经验、技术与方法，重点争取国际合作的智力支持项目。积极与国外保护地建立友好合作备忘录，推进人才交流与营销宣传。

6.3 保障措施

6.3.1 组织领导

组织黄山风景名胜区生物多样性领导小组。生物多样性领导小组以黄山风景名胜区管委会主任为领导，以黄山风景名胜区管委会园林局为核心部门，规划生物多样性管理工作，协调各部门责任，组织实施生物多样性应急任务。

生物多样性领导小组与其他部门积极沟通配合参与。宣传部负责生物多样性保护理念的宣传；纪委监察室负责生物多样性保护监督与应对市民、游客的投诉并反馈至责任部门；规划土地处负责生物多样性保护区划以及配合园林局实施重大项目及工程；经济发展局负责在生物资源的可持续利用先决条件下发展旅游区经济；综合执法局负责该规划的强制性实施。

6.3.2 监督考核

将生物多样性保护目标纳入工作考评体系。通过工作成绩评价制约领导层对不利于生物多样性保护的项目审核与批准，并督促领导层重视生物多样性的保护。制定生物多样性保护责任量表，量化生物多样性保护成果，并提交黄山市人民代表大会表决。对于考核结果应有配套的奖惩措施，定期通报工作进度，反馈有关信息，强化生物多样性保护责任追究，对造成严重后果的予以责任追究，绝不姑息，对重大决策和重大建设项目形成生物多样性保护制约屏障。

实施最严格的监督制度。要求被检查单位或者个人提供相关的文件和资料对开展的工作及行为做出说明，并可以查阅或者复制。生物多样性领导小组应进入违法现场进行拍照、摄像和勘验，责成成员单位或者个人停止违反生物多样性保护相关法律、法规的行为，履行法定义务。对于珍稀野生动物经常出没的区域以及珍稀野生植物生长区域进行监控，并鼓励群众参与到生物多样性保护的工作中，对发现以及举报的违法现象坚决予以追究，并对举报者予以回应，每项案例应形成报告并对社会公开。

6.3.3 资金保障

建立生物多样性保护基金。通过财政拨款形成稳健的资金储备，以保证与生物多样性保护有关的各项工作的顺利开展。增加政府在生物多样性保护的资金投入，用于发展生物多样性保护信息化、景区管理、濒危野生动物保护、天然植被的保护与恢复、外来入侵物种防范。加强生物多样性保护资金的管理，提高资金使用效率。与生物多样性保护有关的项目资金要进行审计与监管，确保项目按计划顺利进行。

建立多方融资渠道。通过调研市场的需求，在黄山生物资源可持续利用的前提下，将生物资源产品市场化。采取 PPP 模式、电子商务模式、P2P 模式等引入社会资金，以及更灵活的政策，充分发挥市场机制在生态资源配置中的作用；定期公开生物多样性保护的建设项目融资意向，引导社会资金向生物多样性保护流转；发挥政府投资主体作用和市场化主导作用，充分利用多渠道商业融资手段，筹集社会资金，扩大资金来源。

6.3.4 科技支撑

加强人才梯队的建设。组织开展生物多样性保护相关的培训，如邀请专家开办讲座，到国际先进生物多样性保护单位参观学习。建设生物多样性保护智库，为黄山生物多样性保护提供精准、专业的技术支撑。多种渠道引进生物多样性保护专业人才，通过提高福利待遇与创造提升机遇，保证良好的生物多样性保护人才发展环境。

  开设科技攻坚课题。针对黄山现阶段存在的问题开展研究，加大科研人力与财力的投入，积极探索生物多样性保护的有效途径与方法。加强硬件投入，生物多样性保护要有各项研究主体的基础数据调查，因此须建立生物多样性监测网络机制，为深入调研提供数据支撑；开展应对气候变化的对策研究，了解黄山生境与气候变化的机理机制，以及动物在气候变化大背景下的迁徙行为，为黄山的珍稀物种与特有物种的保护提供理论支撑；开展林业有害生物防治机理机制的研究，严控外来物种的入侵，对有害生物开展预警与防控作业。

附 表

附表1 黄山风景名胜区主要植被类型统计

植被型	群系	海拔分布	典型物种
山地温性针叶林	黄山松林	900～1 600 m	黄山松、四照花、白檀等
暖性常绿针叶林	马尾松林	900 m以下	马尾松、化香、黄连木等
落叶阔叶林	华东椴林	1 100～1 700 m	华东椴、黄山栎、华箬竹等
	茅栗林	2 000 m以下	茅栗、栲、锥栗等
	黄山玉兰林	700 m以上	黄山玉兰、灯台树、响叶杨等
	黄山栎林	1 000～1 750 m	黄山栎、华东椴、金缕梅等
	枫香林	500～800 m	枫香、大青、淡竹叶等
	绢毛稠李林	950～1 500 m	绢毛稠李、野珠兰、鸭跖草等
	湖北海棠林	1 000～1 500 m	湖北海棠、华箬竹、龙师草等
	小叶白辛树林	400～1 600 m	水马桑、细野麻等
中亚热带常绿阔叶林	细叶青冈林	1 000～1 400 m	细叶青冈、刺芒野古草、岩柃等
	交让木林	600～1 900 m	交让木、山槐、鹅耳枥等
竹林	毛竹林	低海拔广布	毛竹、马尾松、杉木
常绿阔叶灌丛	黄山杜鹃群落	1 000 m以上	虎皮楠、雷公鹅耳枥等
	华箬竹群落	1 000 m以下	合轴荚蒾、山莓等
草丛	拟麦氏草群落	1 500～2 100 m	黄山菊、荠苨等

附表2 黄山风景名胜区国家重点保护高等植物一览

名录	种名	科名	属名	受威胁程度
《国家重点保护野生植物名录》（第一批）（1999）	银杏 Ginkgo biloba	银杏科	银杏属	I
	红豆杉 Taxus wallichiana var. chinensis	红豆杉科	红豆杉属	I
	南方红豆杉 Taxus wallichiana var. mairei	红豆杉科	红豆杉属	I
	榧树 Torreya grandis	红豆杉科	榧树属	II
	鹅掌楸 Liriodendron chinensis	木兰科	鹅掌楸属	II
	连香树 Cercidiphyllum japonicum	连香树科	连香树属	II
	香果树 Emmenopterys henryi	茜草科	香果树属	II
	粗梗水蕨 Ceratopteris pteridoides	水蕨科	水蕨属	II
	水蕨 Ceratopteris thalictroides	水蕨科	水蕨属	II
	金钱松 Pseudolarix amabilis	松科	金钱松属	II
	黄杉 Pseudotsuga sinensis	松科	黄杉属	II
	长序榆 Ulmus elongata	榆科	榆属	II
	大叶榉 Zelkova schneideriana	榆科	榉属	II
	厚朴 Magnolia officinalis	木兰科	木兰属	II
	天竺桂 Cinnamomum japonicum	樟科	樟属	II
	野大豆 Glycine soja	豆科	大豆属	II
	花榈木 Ormosia henryi	豆科	红豆属	II
	金荞麦 Fagopyrum dibotrys	蓼科	荞麦属	II
	黄山梅 Kirengeshoma palmata	虎耳草科	黄山梅属	II
	蛛网萼 Platycrater arguta	虎耳草科	蛛网萼属	II
	秤锤树 Sinojackia xylocarpa	安息香科	秤锤树属	II

附表 3 黄山风景名胜区 IUCN 红色名录和中国物种红色名录植物一览

序号	中文名	学名	科名	IUCN红色名录	中国物种红色名录
1	银杏	*Ginkgo biloba*	银杏科	EN	濒危 EN
2	八角莲	*Dysosma versipellis*	小檗科	VU	易危 VU
3	红豆杉	*Taxus wallichiana* var. *chinensis*	红豆杉科	EN	易危 VU
4	南方红豆杉	*Taxus wallichiana* var. *mairei*	红豆杉科	—	易危 VU
5	榧树	*Torreya grandis*	红豆杉科	LR/NT	易危 VU
6	鹅掌楸	*Liriodendron chinensis*	木兰科	NT	易危 VU
7	厚朴	*Magnolia officinalis*	木兰科	EN	易危 VU
8	草芍药	*Paeonia obovata*	芍药科	—	易危 VU
9	鸡爪槭	*Acer palmatum*	槭树科	—	易危 VU
10	花榈木	*Ormosia henryi*	蝶形花科	—	易危 VU
11	三尖杉	*Cephalotaxus fortunei*	三尖杉科	—	近危 NT
12	粗榧	*Cephalotaxus sinensis*	三尖杉科	—	近危 NT
13	青檀	*Pteroceltis tatarinowii*	榆科	—	近危 NT
14	秀丽槭	*Acer elegantulum*	槭树科	—	近危 NT
15	香果树	*Emmenopterys henryi*	茜草科	—	近危 NT
16	黄山松	*Pinus taiwanensis*	松科	—	近危 NT
17	黄杉	*Pseudotsuga sinensis*	松科	VU	易危 VU
18	金钱松	*Pseudolarix amabilis*	松科	VU	易危 VU
19	铁杉	*Tsuga chinensis*	松科	—	近危 NT
20	长序榆	*Ulmus elongata*	榆科	VU	濒危 EN
21	短萼黄连	*Coptis chinensis*	毛茛科	—	易危 VU
22	天女木兰	*Magnolia sieboldii*	木兰科	LC	易危 VU
23	黄山木兰	*Magnolia cylindrica*	木兰科	VU	易危 VU
24	天目木兰	*Magnolia amoena*	木兰科	VU	易危 VU

序号	中文名	学名	科名	IUCN 红色名录	中国物种红色名录
25	望春玉兰	*Magnolia biondii*	木兰科	—	易危 VU
26	天竺桂	*Cinnamomum japonicum*	樟科	LR/NT	易危 VU
27	连香树	*Cercidiphyllum japonicum*	连香树科	LR/NT	易危 VU
28	紫茎	*Stewartia sinensis*	山茶科	—	近危 NT
29	黄山梅	*Kirengeshoma palmata*	虎耳草科	EN	易危 VU
30	蛛网萼	*Platycrater arguta*	虎耳草科	—	易危 VU
31	黄山花楸	*Sorbus amabilis*	蔷薇科	VU	易危 VU
32	阔叶槭	*Acer amplum*	槭树科	—	近危 NT
33	安徽槭	*Acer anhweiense*	槭树科	—	易危 VU
34	临安槭	*Acer linganense*	槭树科	—	易危 VU
35	毛果槭	*Acer nikoense*	槭树科	—	易危 VU
36	天目槭	*Acer sinopurpurascens*	槭树科	—	易危 VU
37	婺源槭	*Acer wuyuanense*	槭树科	—	易危 VU
38	瘿椒树	*Tapiscia sinensis*	省沽油科	VU	近危 NT
39	明党参	*Changium smyrnioides*	伞形科	—	易危 VU
40	马醉木	*Pieris japonica*	杜鹃花科	—	易危 VU
41	秤锤树	*Sinojackia xylocarpa*	安息香科	VU	濒危 EN
42	白花过路黄	*Lysimachia huitsunae*	报春花科	—	易危 VU
43	安徽羽叶报春	*Primula merrilliana*	报春花科	—	易危 VU
44	黄山龙胆	*Gentiana delicata*	龙胆科	—	易危 VU
45	黄山风毛菊	*Saussurea hwangshanensis*	菊科	—	易危 VU
46	南方兔儿伞	*Syneilesis australis*	菊科	—	易危 VU
47	延龄草	*Trillium tschonoskii*	百合科	—	近危 NT
48	黄精叶钩吻	*Croomia japonica*	百合科	—	易危 VU
49	短穗竹	*Brachystachyum densiflorum*	禾本科	—	易危 VU
50	象鼻兰	*Nothodoritis zhejiangensis*	兰科	—	濒危 EN

序号	中文名	学名	科名	IUCN红色名录	中国物种红色名录
51	钩距虾脊兰	*Calanthe graciliflora*	兰科	—	易危 VU
52	建兰	*Cymbidium ensifolium*	兰科	—	易危 VU
53	长唇羊耳蒜	*Liparis pauliana*	兰科	—	易危 VU
54	广东石豆兰	*Bulbophyllum kwangtungense*	兰科	—	近危 NT
55	蜈蚣兰	*Cleisostoma scolopendrifolium*	兰科	—	近危 NT
56	毛萼山珊瑚	*Galeola lindleyana*	兰科	—	近危 NT
57	鹅毛玉凤花	*Habenaria dentata*	兰科	—	近危 NT
58	二叶兜被兰	*Neottianthe cucullata*	兰科	—	近危 NT
59	舌唇兰	*Platanthera japonica*	兰科	—	近危 NT
60	小舌唇兰	*Platanthera minor*	兰科	—	近危 NT
61	筒距舌唇兰	*Platanthera tipuloides*	兰科	—	近危 NT
62	带唇兰	*Tainia dunnii*	兰科	—	近危 NT
63	小花蜻蜓兰	*Tulotis ussuriensis*	兰科	—	近危 NT
64	白及	*Bletilla striata*	兰科	—	易危 VU
65	虾脊兰	*Calanthe discolor*	兰科	—	易危 VU
66	蕙兰	*Cymbidium faberi*	兰科	—	易危 VU
67	春兰	*Cymbidium goeringii*	兰科	—	易危 VU
68	扇脉杓兰	*Cypripedium japonicum*	兰科	—	易危 VU
69	独蒜兰	*Pleione bulbocodioides*	兰科	—	易危 VU
70	无柱兰	*Amitostigma gracile*	兰科	—	近危 NT
71	金线兰	*Anoectochilus roxburghii*	兰科	—	近危 NT
72	金兰	*Cephalanthera falcata*	兰科	—	近危 NT

附表4 黄山风景名胜区内列入 CITES 附录 II 植物一览

序号	中文名	学名	科名	属名
1	白及	*Bletilla striata*	兰科	白及属
2	虾脊兰	*Calanthe discolor*	兰科	虾脊兰属
3	钩距虾脊兰	*Calanthe graciliflora*	兰科	虾脊兰属
4	反瓣虾脊兰	*Calanthe reflexa*	兰科	虾脊兰属
5	蕙兰	*Cymbidium faberi*	兰科	兰属
6	春兰	*Cymbidium goeringii*	兰科	兰属
7	扇脉杓兰	*Cypripedium japonicum*	兰科	杓兰属
8	独蒜兰	*Pleione bulbocodioides*	兰科	独蒜兰属
9	无柱兰	*Amitostigma gracile*	兰科	无柱兰属
10	金线兰	*Anoectochilus roxburghii*	兰科	开唇兰属
11	金兰	*Cephalanthera falcata*	兰科	头蕊兰属
12	斑叶兰	*Goodyera schlechtendaliana*	兰科	斑叶兰属
13	绶草	*Spiranthes sinensis*	兰科	绶草属
14	广东石豆兰	*Bulbophyllum kwangtungense*	兰科	石豆兰属
15	蜈蚣兰	*Cleisostoma scolopendrifolium*	兰科	隔距兰属
16	建兰	*Cymbidium ensifolium*	兰科	兰属
17	毛萼山珊瑚	*Galeola lindleyana*	兰科	山珊瑚属
18	鹅毛玉凤花	*Habenaria dentata*	兰科	玉凤花属
19	长唇羊耳蒜	*Liparis pauliana*	兰科	羊耳蒜属
20	二叶兜被兰	*Neottianthe cucullata*	兰科	兜被兰属
21	象鼻兰	*Nothodoritis zhejiangensis*	兰科	象鼻兰属
22	舌唇兰	*Platanthera japonica*	兰科	舌唇兰属
23	筒距舌唇兰	*Platanthera tipuloides*	兰科	舌唇兰属
24	小舌唇兰	*Platanthera minor*	兰科	舌唇兰属
25	带唇兰	*Tainia dunnii*	兰科	带唇兰属
26	小花蜻蜓兰	*Tulotis ussuriensis*	兰科	蜻蜓兰属

附表 5　黄山风景名胜区国家重点保护动物一览

序号	种名	国家重点保护级别	备注
1	白颈长尾雉 Syrmaticus ellioti	I	典型的东洋界华中区东部丘陵平原亚区种类
2	白鹳 Ciconia ciconia	I	分布于欧洲、非洲西北部、亚洲西南部和非洲南部，长途迁徙性鸟类
3	云豹 Neofelis nebulosa	I	世界性保护物种
4	金钱豹 Panthera pardus	I	从低山、丘陵至高山森林、灌丛均有分布
5	黑麂 Muntiacus crinifrons	I	《濒危野生动植物种国际贸易公约》（CITES）附录I物种，世界自然保护联盟（IUCN）将其濒危等级列为易危
6	梅花鹿 Cervus nippon	I	《中国物种红色名录》：EN，IUCN 红色名录：LC
7	大鲵 Andrias davidianus	I	孑遗物种
8	鸳鸯 Aix galericulata	II	观赏鸟类
9	白鹇 Lophura nycthemera	II	观赏鸟类
10	勺鸡 Pucrasia macrolopha	II	观赏鸟类
11	黑鸢 Milvus migrans	II	猛禽
12	赤腹鹰 Accipiter soloensis	II	猛禽
13	雀鹰 Accipiter nisus	II	猛禽
14	普通鵟 Buteo buteo	II	猛禽
15	毛脚鵟 Buteo lagopus	II	猛禽
16	乌雕 Aquila clanga	II	猛禽
17	红隼 Falco tinnunculus	II	猛禽
18	短尾猴 Macaca arctoides	II	地栖性较强
19	猕猴 Macaca mulatta	II	乱捕滥猎是猕猴致危的主要因素
20	穿山甲 Manis pentadactyla	I	穿山甲野外数量稀少，在中国，禁止私人捕杀和食用
21	豺 Cuonalpinus	II	列入 IUCN 红色名录
22	黑熊 Ursus thibetanus	II	药用动物，遭受捕杀
23	大灵猫 Viverra zibetha	II	药用
24	小灵猫 Viverricula indica	II	列入《濒危野生动植物种国际贸易公约》（CITES）附录III和世界自然保护联盟（IUCN）红色名录
25	獐 Hydropotes inermis	II	珍贵药用动物
26	鬣羚 Capricornis sumatraensis	II	《中国物种红色名录》和《中国濒危动物红皮书》将其列为易危级（VU）

附表6 黄山风景名胜区的经济价值分类构成和估算结果一览　　　　单位：10^8 元/a

价值分类	构成内容	价值估算		合计
直接开发价值	旅游收入	28.70		35.12
	科研价值	0.42		
	文化价值	2.00		
	产品价值	4.00		
生态保护价值	生态效益价值	涵养水源	4.56	208.54
		保持土壤	3.00	
		调节气候	2.71	
		控制灾害	0.21	
		净化环境	0.68	
		小计	11.16	
	生物资源价值	自然生物资源	3.09	
		商业生物资源	193.29	
		生物多样性服务	1.00	
		小计	197.38	
未来存在价值	—	388.67		388.67

附表7 黄山风景名胜区旅游环境容量测算结果　　　　单位：人/d

容量＼景区	温泉景区	云谷景区	玉屏景区	北海景区	天海景区	西海景区	钓桥景区	东海景区	合计
夏半年日极限容量	2 095	5 295	7 061	6 110	9 579	5 254	17 302	12 150	64 846
冬半年日极限容量	1 348	2 480	4 512	4 926	8 040	3 427	14 604	9 323	48 660
非高峰日合理容量	962	1 766	3 203	3 873	5 725	2 454	10 499	6 325	34 807

参考文献

[1] 王祥荣，钱阳平. 黄山风景名胜区资源环境评估与基础数据库建构[M]. 北京：科学出版社，2018.

[2] 林清贤，钱阳平，张丽荣，等. 黄山鸟类[M]. 北京：中国环境出版集团，2018.

[3] 刘春生. 九龙山自然保护区珍稀濒危植物黄山木兰种群生态学研究[D]. 杭州：浙江师范大学，2010.

[4] 黄山风景名胜区. 关于印发《黄山风景名胜区管委会权责清单制度建设和两个服务清单工作实施方案》的通知. http://hsgwh.huangshan.gov.cn/BranchOpennessContent/show/840280.html.

[5] 毕淑峰. 黄山风景名胜区的珍稀植物资源[J].国土与自然资源研究，2004（4）：95-96.

[6] 陈文豪. 安徽石台云豹频繁出现的原因初探[J]. 安徽林业科技，2003（4）：11-12.

[7] 汤宛地. 松材线虫病入侵黄山风景名胜区的风险性评估[D]. 北京：北京林业大学，2008.

[8] 吴磊. 黄山风景名胜区外来植物入侵风险的研究[D]. 合肥：安徽农业大学，2012.

[9] 郑伟成，朱爱军，张方纲，等. 九龙山自然保护区国家重点保护野生植物优先保护序列研究[J]. 浙江林业科技，2012，32（6）：39-43.

[10] 孙语圣. 神灵与宗教：明清时期徽州与淮北民间信仰之比较[J]. 社会科学，2018（3）：162-171.

[11] 李群根. 徽派建筑的美学艺术[J]. 四川建材，2018，44（3）：26-28.

[12] 汪珺. 浅谈新安医学[J]. 教育教学在线，2018（20）：217-218.

[13] 方德国. 徽菜的烹制与合理营养研究[J]. 楚雄示范学院学报，2018（33）：22-25.

[14] 黄山市统计局. 2017黄山市统计年鉴. 2018：388-389.

[15] 肖娟. 黄山市旅游业对城镇化发展的负效应研究[J]. 湖南商学院学报，2013，20（5）：62-66.

[16] 耿雪梅，邱瑛. 黄山风景名胜区发展现状及对策研究[J]. 对外经贸，2018（284）：119-120.

[17] 故宫博物院. https://www.dpm.org.cn/Home.html.

[18] 王群，陆林，杨兴柱. 缺水型山岳景区水资源安全影响因素分析——以黄山风景名胜区为例[J]. 干旱区资源与环境，2014，28（11）：48-53.

[19] 肖娟. 黄山市旅游业对城镇化发展的负效应研究[J]. 湖南商学院学报（双月刊），2013，20(5)：62-66.

[20] 刘正威. 黄山世界文化与自然双遗产的可持续发展研究[D]. 北京：中国地质大学，2013.

[21] 张金泉. 基于CVM的黄山旅游资源非使用价值评估研究[D]. 上海：上海师范大学，2007.

[22] 章尚正. 徽州文化生态保护与利用的四维发展[J]. 黄山学院学报，2010，12（1）：1-6.

[23] 陆林. 皖南旅游区布局研究[J]. 地理科学，1995（1）：88-95+100.

[24] 吴丽蓉. 徽州文化旅游深度开发与对策研究[D]. 芜湖：安徽师范大学，2013：32-33.

[25] OECD. The economic appraisal of environmental protects and policies：apractical guide [M]. Paris：OEDE，1995.

[26] 丁晖，徐海根. 生物物种资源的保护和利用价值评估——以江苏省为例[J]. 生态与农村环境学报，2010，26（5）：454-460.

[27] 吴火和. 森林生物多样性资产价值评估研究[D]. 福州：福建农林大学，2006：4.

[28] 李文华，等. 生态系统服务功能价值评估的理论、方法与应用[M]. 北京：中国人民大学出版社，2008.

[29] 谢高地，张钇锂，鲁春霞，等. 中国自然草地生态系统服务价值[J]. 自然资源学报，2001，16（1）：46-53.

[30] 马克平，钱迎倩，王晨. 生物多样性研究的现状与发展趋势[J]. 基础科学，1995（1）：27-30.

[31] 吴征镒，彭华. 生物资源的合理开发利用和生物多样性的有效保护[J]. 世界科技研究与发展，1996，18（1）：24-30.

[32] 郭中伟，李典谟. 生物多样性经济价值评估的基本方法[J]. 生物多样性，1999，7（1）：60-67.

[33] 董冬，周志翔，何云核，等. 基于游客支付意愿的古树名木资源保护经济价值评估——以安徽省九华山风景区为例[J]. 长江流域资源与环境，2011，20（11）：1334-1340.

[34] 凌田心. 黄山市中药资源调查及开发利用[J]. 基层中药杂志，2000，14（5）：40-41.

[35] 黄山茶叶有望突破10亿元大关. 中国茶网. http://ahcha.anhuinews.com/system/2012/05/--23/004974321.shtml.

[36] 刘璇. 基于生物多样性价值的中国驰名商标生物存在价值评估[D]. 北京：北京林业大学，2013.

[37] 宋娓娓. 基于徽州地域文化背景的旅游产品设计研究[D]. 芜湖：安徽工程大学，2012.

[38] 胡生新. 生态系统服务功能及其价值评价研究概述[J]. 甘肃科技，2009，25（12）：86-88，101.

[39] 葛惠. 黄山风景名胜区休闲旅游发展研究[D]. 南宁：广西大学，2015.

[40] 刘正威. 黄山世界文化与自然双遗产的可持续发展研究[D]. 北京：中国地质大学，2013.

[41] 张金泉. 基于CVM的黄山旅游资源非使用价值评估研究[D]. 上海：上海师范大学，2007.

[42] 王群. 黄山风景名胜区旅游与水环境协调发展研究[D]. 芜湖：安徽师范大学，2005.

[43] Buckley R. A framework for ecotourism[J]. Annals of Tourism Research，1994，21（3）：661-665.

[44] 章锦河，张捷. 旅游生态足迹模型及黄山市实证分析[J]. 地理学报，2004（5）：763-771.

[45] 徐艳，朱生东，聂祝兰. 徽文化旅游与生态旅游融合发展研究[J]. 遵义师范学院学报，2013，

15（1）：16-19.

[46] 肖娟. 黄山市旅游业对城镇化发展的负效应研究[J]. 湖南商学院学报，2013，20（5）：62-66.

[47] 笪忠敏，何静，王浩，等. 基于旅游者需求视角下的我国智慧景区运营模式研究——以安徽黄山风景名胜区为例[J]. 旅游纵览（下半月），2015（2）：205-207.

[48] 孙春华. 山地风景区旅游环境承载力及其调控系统研究[D]. 济南：山东师范大学，2002.

[49] 巩劼，陆林，晋秀龙，等. 黄山风景名胜区旅游干扰对植物群落及其土壤性质的影响[J]. 生态学报，2009，29（5）：2239-2251.

[50] 王群，陆林，杨兴柱. 缺水型山岳景区水资源安全影响因素分析——以黄山风景名胜区为例[J]. 干旱区资源与环境，2014，28（11）：48-53.

[51] 田艳. 黄山风景名胜区生态风险分析与评价研究[D]. 芜湖：安徽师范大学，2010.

[52] 崔凤军. 论旅游环境承载力——持续发展旅游的判据之一[J]. 经济地理，1995，15（1）：105-109.

[53] 胡炳清. 旅游环境容量计算方法[J]. 环境科学研究，1995，8（3）：20-24.

[54] 刘会平，唐晓春，蔡靖芳，等. 武汉东湖风景区旅游环境容量初步研究[J]. 长江流域资源环境，2001，10（3）：230-235.

[55] 郎咏梅，孙洪涛，田家怡，等. 崂山风景区旅游环境容量研究[J]. 中国人口·资源与环境，2006，16（4）：99-102.

[56] 卢学英，蒋宁，胡利军. 黄山风景名胜区旅游环境承载力研究[J]. 河南工程学院学报（社会科学版），2018，33（1）：25-28.

[57] 孙春华. 山地风景区旅游环境承载力及其调控系统研究[D]. 济南：山东师范大学，2002.

[58] 罗婷婷. 黄山风景名胜区社区问题与社区规划研究[D]. 北京：清华大学，2004.

[59] 吴丽媛，陈传明，侯雨峰. 武夷山风景名胜区旅游环境容量研究[J]. 资源开发与市场，2016，32（1）：108-111.

[60] 崔凤军. 泰山旅游环境承载力及其时空分异特征与利用强度研究[J]. 地理研究，1997，16（4）：47-54.

[61] 寇文波，李德云，秦愿，等. 国内旅游容量研究述评[J]. 首都师范大学学报（自然科学版），2017，38（5）：56-61.

Action Plans for Biodiversity Conservation within the Huangshan Scenic Area
（2018—2030）

Preface

Biodiversity is a necessary condition for human survival and development, and the material basis for maintaining national ecological security. Since the international community reached an agreement and signed the *Convention on Biological Diversity* in 1992, and with the joint efforts of all parties, positive progress has been made in the conservation of the world's biodiversity. China is a country that has some of the richest biodiversity in the world. The Chinese government attaches great importance to the protection of biodiversity, and regards it as an important part of the construction of ecological civilization and as an effective means to promote high-quality development. In recent years, under the guidance of Xi Jinping's concepts concerning ecological civilization, the Chinese government has made unremitting efforts to speed up the mainstreaming process of biodiversity conservation, improve social participation and public awareness, and actively contribute to global biodiversity conservation. However, we should also be aware of the threats to biodiversity in China, the severity of which rank highly in the world. In fact, the decline of biodiversity in China is accelerating, and this general trend has not yet been effectively curbed.

The Mount Huangshan Scenic Spot (hereinafter referred to as Mount Huangshan) is located in Huangshan City, Anhui Province. It is recognized as a national civilized and 5A level scenic spot, and was listed among the first batch of key national scenic spots by the State Council in 1982. Furthermore, Mount Huangshan was listed on the *World Cultural and Natural Heritage List* in 1990, became a Global Geopark in 2004, and joined the World Network of Biosphere Reserves in 2018. Located in the northern margins of the central subtropical zone, this scenic spot is a refuge for animals and plants in the Quaternary Glacier Period, and it has a unique geographical location, rolling mountains, diverse landforms, and an elevation difference of more than 1,600 meters. As a result, Mount Huangshan is rich in biodiversity, with 2,385 species of higher plants and 417 species of vertebrates. Although it only accounts for 0.044% of China's land area, it has 6.92% of the plant species and 9.55% of the animal species in China. In recent years, the Mount Huangshan scenic spot has significantly improved its biodiversity conservation capacity by innovating a system of "alternating rest" of the scenic spots, strengthening the protection of the ancient and famous trees, strictly carrying out plant quarantines, strengthening fire prevention measures, and actively protecting the water resources.

In order to implement the relevant requirements and mandates of the state, provinces, and municipalities, further strengthen the protection and management of biodiversity in scenic spots, and effectively address the new problems and challenges facing biodiversity conservation, the Environmental Planning Institute of the Ministry of Ecology and Environment, as entrusted by the Landscape Bureau of the Scenic Area Management Committee, organized a technical team to compile its *Action Plan for Biodiversity Conservation in the Mount Huangshan Scenic Area* (2018-2030) (hereinafter referred to as the Action Plan). In May 2019, the Huangshan Municipal People's Government approved and agreed to implement the Action Plan, striving to make Mount Huangshan a model for biodiversity conservation in mountain scenic spots throughout the country by strengthening the development of biodiversity conservation measures there.

As the first biodiversity conservation action plan for a scenic spot in China, this book estimates the environmental capacity of scenic spots and the carrying capacity of the tourism environment by using an instantaneous calculation method of the scenic spots' space, and by comprehensively combing the general situation, characteristics, cultural value, value evaluation, current situation, and progress of the conservation and management of scenic spots. In addition, the carrying capacity of the tourism environment is evaluated. By evaluating and identifying the importance of the function of biodiversity conservation and the habitat space of protected species, the important areas for biodiversity conservation are delineated. Based on an in-depth analysis of the problems facing biodiversity conservation, the objectives and main tasks for biodiversity conservation in scenic spots were put forward, and 10 priority areas and 30 priority actions were systematically planned.

Thanks to Party Secretary Lu Jun, Academician Wang Jinnan, Prof. Wang Xiahui (Chinese Academy of Environmental Planning, CAEP), Prof. Zhang Fengchun, Prof. Li Junsheng (Chinese Researrch Academy of Environmental Sciences, CRAES), Prof. Min Qingwen, Prof. Zhong Linsheng (Institute of Geographic Sciences and Natural Resources Research), Prof. Xue Dayuan (Minzu University of China), Prof. Wang Jianzhong, Prof. Wang Fengjun (Beijing Forestry University), Prof. Zeng Weihua (Beijing Normal

University) and more notable scholars and experts for their valuable advice and help in this book.

Thanks to the Landscape Bureau of the Mount Huangshan Scenic Area Management Committee for their support and assistance.

Thanks to all the authors in the references for their stimulating insights, which have inspired the writing of this book. We apologize for any omissions of quotations there may be in the book.

Thanks to China Environment Publishing Group for supporting the publication of this book.

This book is suitable for experts and scholars engaged in the field of biodiversity conservation and management in scenic spots, managers and decision makers of the relevant government departments, graduate students in biodiversity-related fields, and other persons who are interested in it for reading and reference. If there are any mistakes or inappropriate elements, we would like to invite all the experts, scholars, and readers to contribute their valuable comments.

The Author

In Beijng, July, 2020

1 Introduction

1.1 Overview of the Mount Huangshan Scenic Area

The Mount Huangshan scenic area is located in the administrative area of Huangshan District, Huangshan City, Anhui Province (Figure 1-1). It was listed among the first batch of National Parks in China by the State Council in 1982, and was also listed on the National Civilized Scenery Tourist Area Demonstration Sites and the National AAAAA Level Tourist Attractions. Moreover, Huangshan was identified as a World Cultural and Natural Heritage Site in 1990, became a World Geopark in 2004, and joined the World Network of Biosphere Reserves in 2018.

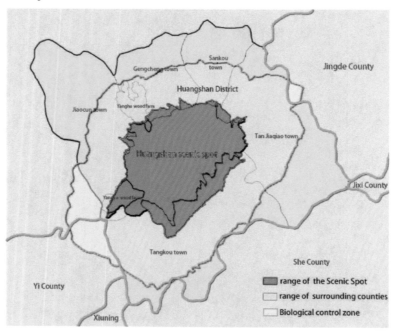

Figure 1-1　The Range of the Plan for Biodiversity Conservation within the Huangshan Scenic Area

In December 1988, the Mount Huangshan Scenic Area Management Committee was established, with 16 subsidiary organizations, including 8 county-level organizations: the Discipline Committee Office (Supervision Office, Economic Audit Bureau), Political Section, Bureau of Landscaping, Land Management Division (Huangshan Geopark Administration), Bureau of Economic Development (Bureau of Finance), Comprehensive Enforcement Bureau (Comprehensive Management Office, Huangshan Scenic Area Branch of Huangshan Comprehensive Enforcement Bureau of Tourism Management), and Public Security Bureau. It also has 6 assisting county-level organizations: the Party Committee, Labour Union, the Communist Youth League (managed by the Political Section), Propaganda Department (Civilization Office), People's Armed Forces Department, Market Supervision Administration. In addition, it has 2 division-level organizations: the Bureau of Roads and Traffic and the Huashan Mysterious Grottoes—Jian River Scenic Area Management Office. The mayor of Huangshan concurrently serves as the director of the management committee, which is responsible for the protection, utilization, and unified management of the scenic spots under the leadership of the Huangshan Municipal People's government.

Taking tourism as its pillar industry, the Huangshan scenic area has seen a significant change in its tourism business income in the past decade. In 2017, the Huangshan scenic area received a total of 3.3687 million tourists, a 50.4% increase compared with ten years ago (2.24 million tourists in 2008). However, the economic development of Tangkou town, Tanjiaqiao town, Sankou town, Geng town, Jiaocun town and the state-owned Yanghu Forest Farm adjacent to the Huangshan scenic area is very uneven. In addition, the residents' income from "five towns and one farm" mainly comes from agriculture, forestry, and tourism.

1.2 The Origin of BSAP in the Huangshan Scenic Area

Mount Huangshan has become a well-known scenic area in the world because of its unique natural and cultural landscape resources and rich biodiversity, including majestic peaks, strange rock formations, ancient trees, famous pines, the "cloud sea", and the hot spring and glacier remains. These features have distinguished Mount Huang as one of the 108 biodiversity distribution centers in the world, according to the International Union for the Conservation of Nature, and as one of the 33 priority areas for biodiversity conservation in China (Mount Huangshan—Mount Huaiyu priority areas for biodiversity conservation). Mount Huangshan is an important peripheral-mountain water source of the Qiantang River, and the biodiversity of Mount Huangshan and its surrounding communities will directly affect the downstream water quality of the Qiandao River and the Thousand Island Lake (a freshwater lake), and the well-being of the people living in Zhejiang province. Therefore, the quality of the biodiversity around Mount Huangshan has an important impact on regional ecological security and also provides economic benefits.

As the first scenic area in China to formulate an action plan for biodiversity conservation, through scientific deployment and the systematic implementation of "biodiversity-protection" related fields and actions, the Huangshan scenic area can further highlight its important position and the comprehensive benefits of its biodiversity, continuously improving public education and the tourism experience there, while also enhancing the fair distribution of benefits and community happiness.

1.3 The Relationship between Mount Huangshan and NBSAP

The *China Biodiversity Conservation Strategy and Action Plan (2011-2030)* (referred to as NBSAP) was deliberated on and passed at the 126th executive meeting of the State Council, and was issued on September 17, 2010 as the general outline for China's biodiversity conservation. Based on full integration with the NBSAP, the *Strategy and Action Plan for Biodiversity Conservation in the Huangshan Scenic Area (2018-2030)* has reflected and innovated on the NBSAP in accordance with the actual situation on the ground.

1.3.1 The Consistent Overall Objectives

The NBSAP requires implementing the Scientific Outlook on Development, the overall planning of biodiversity protection and economic development, with the goal of protecting and sustainably utilizing biodiversity, while also providing for the fair and reasonable sharing of benefits generated by the utilization of genetic resources. The phased goals for national biodiversity protection have 2015, 2020, and 2030 as the time nodes.

The *Strategy and Action Plan for Biodiversity Conservation in the Huangshan Scenic Area (2018-2030)* aims at improving and maintaining the service quality of the scenic area ecosystem and the area of protected space, while strictly abiding by the bottom line of biodiversity protection, and coordinating and unifying the biodiversity protection and the economic and social development of the scenic area, all with the priority action of biodiversity protection as the starting point. These are consistent with the overall objectives of the NBSAP.

1.3.2　The Consistency of Dividing Important Areas

According to the natural, social, and economic conditions, and to the distribution characteristics of the natural resources and primary protected objects, the NBSAP divides the whole country into 8 natural areas (i.e. the northeast mountain plain area, Mongolian new plateau desert area, north China plain Loess Plateau area, alpine region of the Qinghai Tibetan Plateau, southwest mountain and canyon area, middle, south and western hilly areas, east and central China hilly plain area, and south China low-mountain hilly area). Based on this division, the NBSAP has comprehensively considered the representativeness, uniqueness, special ecological functions, and species diversity, degree of rarity and endangerment, degree of threat, regional representativeness, economic use, scientific research value, distribution data acquisition, and other ecosystem-type factors, thereby designating 35 priority areas for biodiversity protection. The Huangshan scenic area is located in the Mount Huangshan—Mount Huaiyu priority area for biodiversity protection in the east and central China hilly plain area.

In the *Strategy and Action Plan for Biodiversity Conservation in the Huangshan Scenic Area (2018-2030)*, the integrity of the biological ecosystem, the habitat of the special species and the distribution of important genetic resources in the Huangshan scenic area are the main reference factors used to evaluate and identify the important areas of biodiversity protection in Mount Huangshan on two levels. These important areas are divided into four categories: the biodiversity protection barrier area, high adaptability area for protected species, landscape viewing area, and vulnerable area for natural disasters.

1.3.3　The Consistency in the Priority Areas and Actions

While the NBSAP has identified 10 priority areas and 30 priority actions for China's biodiversity conservation by 2030, the *Strategy and Action Plan for Biodiversity Conservation in the Huangshan Scenic Area (2018-2030)* has also deployed 10 priority areas and 30 priority actions.

Among them, strengthening the investigation and monitoring of biodiversity, scientifically carrying out local and off-site conservation for biodiversity, promoting the sustainable development and utilization of biological resources, strengthening the management of invasive alien species, improving the ability to cope with climate change, improving the policies and systems related to biodiversity protection, and promoting public awareness and education are the common goals of both the NBSAP and the *Strategy and Action Plan for Biodiversity Conservation in the Huangshan Scenic Area （2018-2030）*.

In addition, according to the basic capacity, biodiversity characteristics, and development needs of the Huangshan scenic area, the *Strategy and Action Plan for Biodiversity Conservation in the Huangshan Scenic Area （2018-2030）* puts forward a series of preliminary plans based on the NBSAP, such as biodiversity information management, special protection for rare and endangered endemic species, ecosystem restoration, and promotion. It has coordinated the development of the communities surrounding the scenic area, and has designed 13 corresponding priority actions.

2 Basic BSAP Research on the Mount Huangshan Scenic Spot

2.1 Biodiversity of the Huangshan Scenic Area

Located in the northern margins of the central subtropical zone, this scenic spot is a refuge for animals and plants in the Quaternary Glacier Period, and has a unique geographical location, rolling mountains, diverse landforms, and an elevation difference of more than 1,600 meters. As a result, the Huangshan scenic area is rich in biodiversity, with 2,385 species of higher plants and 417 species of vertebrates. Although it only accounts for 0.044% of China's land area, it has 6.92% of the plant species and 9.55% of the animal species in China.

2.1.1 Ecosystem Diversity

The forest ecosystem area in Mount Huangshan is the widest and largest, accounting for 89.48% of the total area, while the second-largest ecosystems (such as bare rock) account for 6.50% of the total area, which are mainly scattered in the hot springs, Yungu, and Diaoqiao management areas. The grassland ecosystem accounts for about 0.40% of the total scenic area, and is located mainly in the Fuxi management zone. The areas of the wetland ecosystem and urban ecosystem are small, accounting for less than 0.2% of the scenic area, and they are lightly scattered in the southeastern part of the scenic spot (Figrue 2-1).

Mount Huangshan is rich in forest resources, and its forest coverage is about 98.29%. According to an analysis of the remote sensing data, the total area of the forest ecosystem in the scenic area is about 143.7 square kilometers, accounting for 89.5% of the total scenic area (Figrue 2-2). The forest types mainly consist of evergreen coniferous forest, with Pinus taiwanensis as the constructive species (120.74 square kilometers), evergreen broad-leaved forest, with Fagaceae and Lauraceae trees as the constructive species (3.37

square kilometers), deciduous broad-leaved forest (9.42 square kilometers), and deciduous shrub forest (9.30 square kilometers).

Figrue 2-1　Ecosystem Types in Huangshan Scenic Area

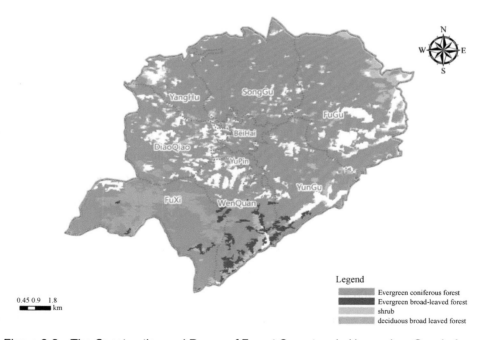

Figrue 2-2　The Construction and Range of Forest Cosystem in Huangshan Scenic Area

2.1.2 Diversity of the Plant Resources

There are many kinds of plants and complex plant communities in Mount Huangshan. There are 178 families, 776 genera, and 1,924 species (including varieties, subspecies and variants) of naturally distributed vascular plants, which are listed in Table 2-1. Among them, there are 161 species in 66 genera and 37 families of ferns, 29 species in 22 genera and 6 families of gymnosperms, and 1,724 species in 688 genera and 135 families of angiosperms. Among the angiosperms, there are 119 families, 555 genera, and 1,455 species of dicotyledons, and 16 families, 133 genera, and 269 species of monocotyledons[①].

Table 2-1 Statistics of the Vascular Plant Composition in the Mount Huangshan Scenic Area

Plant Type	Number of Families	Proportion/%	Number of genera	Proportion/%	Number of Species	Proportion/%
Ferns	37	20.79	66	8.51	161	8.37
Gymnosperms	6	3.37	22	2.83	39	22.03
Angiosperms	135	75.84	688	88.66	1,724	89.60
Total	178	100	776	100	1,924	100.00

The results of the analysis of the geographical elements of the families, genera, and species of seed plants show that the families of Mount Huangshan are mainly distribute in the pantropics (48.5%), among which 59.6% are distributed in the tropical geographical regions. The genera are mainly distributed in the northern temperate zone (23.7%), among which 61.7% are distributed in the temperate geographical regions. The species are mainly endemic to China, among which 33.1% are distributed in the temperate geographical regions (Column 2-1).

> Column 2-1 Diversity of Plants in the Mount Huangshan Scenic Area
>
> According to Wu Zhengyi's *Distribution Types of Seed Plants in China*, there are 14 distribution types of seed plants in the Mount Huangshan scenic spot different from those in Central Asia, and 55 genera belonging to the worldwide distribution type, accounting for

---

① Wang Xiangrong, Qian Yangping. Assessment of the Resources and Environment and Construction of the Basic Database in the Huangshan Scenic Spot [M]. Beijing: Science Press, 2018.

8.40% of the total genera. Among these, *Polygonum*, *Clematis*, *Rubus*, *Viola*, *Pearl Vegetable*, and *Poa* all contain more than ten species. There are 209 genera distributed in the tropical geographical regions, accounting for 31.91% of the total genera.

The greatest number of genera and their variants(in total 99 genera) are distributed in the pantropical geographical regions, accounting for 47.37% of the total genera distributed in the tropical geographical regions. There are 362 genera distributed in the temperate geographical regions, accounting for 55.27% of the total genera.

Among these, those distributed in the northern temperate geographical regions are dominant, with 137 genera accounting for 37.85% of the total genera in the temperate geographical regions. Among the genera distributed in the northern temperate geographical regions, *Quercus*, *Acer*, and *Anthracene* are the common species of forest vegetation in the region, followed by those distributed in East Asia, with 109 genera, accounting for 30.11% of the total genera distributed in the temperate geographical region. In addition, 22 genera are endemic to China. Among them, there are 2 species in the genus *Chimonanthus*, and 1 species in the other genera. There are also numerous singular genera, such as *Codonopsis pilosula* and *Eucommia*.

2.1.3　Diversity of the Animal Resources

Mount Huangshan, with its many peaks and valleys and dense vegetation, provides an ideal habitat and breeding place for wildlife. Therefore, the region has abundant wildlife resources. The dominant species of wild animals in Mount Huangshan are Oriental species (i.e. species originating from the tropical and subtropical regions). There are 8 orders, 22 families, and 73 species of mammals, accounting for 76% of the 96 total species of mammals in Anhui Province; 17 orders, 55 families, and 244 species of birds, accounting for 60% of the total bird species in Anhui Province; as well as 29 species of amphibians in 8 families, 2 orders and 54 species of reptiles in 9 families, 2 orders[①].

2.1.4　Diversity of the Genetic Resources

（1） Various Species of the Forest Genetic Resources

Mount Huangshan has high forest coverage, large forest reserves, diverse forest types, and abundant forest resources. Its forest genetic resources mainly include: *Pinus taiwanensis*, *Cupressus funebris*, *Liriodendron chinense*, *Acer buergerianum*, and

---

① Wang Xiangrong and Qian Yangping et al. Assessment of the Resources and Environment and Construction of the Basic Database of the Huangshan Scenic Spot [M]. Beijing: Science Press, 2018.

Phyllostachys heterocycle cv. *Pubescens*, among others, which are listed in Table 2-2.

Table 2-2 List of the Representative Forest Genetic Resources

Resource Type	Species Name	Family Name
Forest Genetic Resources	*Cupressus funebris*	Cupressaceae
	Pinus taiwanensis	Pinaceae
	Liriodendron chinense	Magnoliaceae
	Acer buergerianum	Aceraceae
	Cinnamomum bodinieri	Lauraceae
	Phyllostachys heterocycle cv. *Pubescens*	Gramineae

(2) Abundant Medicinal Genetic Resources

There are many kinds of plants and abundant medicinal genetic resources in Mount Huangshan. According to the results of the survey, there are more than 700 species of medicinal plants in Huangshan Mountain, accounting for 38.9% of the total plant species. The important medicinal genetic resources include: *star anise*, *Akebia trifoliate*, and *Magnolia officinalis*, among others, which are listed in Table 2-3.

Table 2-3 List of the Representative Medicinal Genetic Resources

Resource Type	Species Name	Family Name
Medicinal Genetic Resources	*Dysosma versipellis*	Berberidaceae
	Coptis chinensis	Ranunculaceae
	Akebia trifoliata	Lardizabalaceae
	Magnilia officinalis	Magnoliaceae
	Atractylodes macrocephala	Asteraceae
	Changium sntyrniodies	Campanulaceae

(3) Unique Floral Genetic Resources

Mount Huangshan has beautiful scenery, precipitous peaks, ancient trees, and abundant floral genetic resources. The unique geographical and geological conditions and flora there have formed the unique floral composition of Mount Huangshan, including: *Kirengeshoma Yatabe*, *Rhododendron anhwiense*, *Magnolia siebodii*, *Enkianthus quinque*, and *Dendrobenthamia japonica* var. *chinensis*, among others, which are listed in Table 2-4.

Table 2-4 List of the Representative Floral Genetic Resources

Resource Type	Species Name	Family Name
Floral Genetic Resources	*Kirengeshoma Yatabe*	Saxifragaceae
	Magnolia siebodii	Magnoliaceae
	Rhododendron anhwiense	Ericaceae
	Brachystachyum densiflorum	Gramineae
	Enkianthus quinque	Ericaceae
	Dendrobenthamia japonica var. *chinensis*	Cornaceae
	Cymbidium goeringii	Orchidaceae
	Rosa laevigata	Rosaceae

(4) Agricultural Genetic Resources

Mount Huangshan has abundant agricultural genetic resources, including crops, livestock, and poultry for production. Mount Huangshan also has abundant tea resources and is a famous tea producing area in China. There are many famous teas, including Huangshan Maofeng, Taiping Houkui, Keemun Black Tea, and many others. The region's specialty tea, *Camellia sinensis* (Huangshanzhong) belongs to Camellia of the Camellia family, and it originated in the Huangshan area of Shexian County, Anhui Province. It is widely planted because of its strong survivability and high yield (Column 2-2).

Column 2-2   Overview of the Tea Resources in Mount Huangshan

The geographical distinctions of the teas in Huangshan are Keemun Black Tea (Brand: Qishan) and Huangshan Maofeng (Brand: Caoxi), and the products protected by the origin area are Taiping Houkui, which is produced in Sankou Town of Huangshan District. In 2005, the General Administration of Quality Supervision, Inspection, and Quarantine approved 6 enterprises in Huangshan District, including Xinming Monkey Village Tea Farm and Huangshan Zhongming Tea Industry Co., Ltd., allowing them to use the special label of "Taiping Monkey Queen Tea" for their original products. By officially awarding this label to these six enterprises, the government of Huangshan District ensured the quality of the famous and excellent tea products in Huangshan, curbed the trend of counterfeit products, and greatly improved the image of the products and the value of the enterprises' intangible assets.

2.2 Characteristics of the Biodiversity in the Huangshan Scenic Area

2.2.1 Various and Fragile Ecosystems

Huangshan Scenic Area is located in the intersection between the subtropical and temperate zones. It is the watershed of the Yangtze River and Qiantang River in southern Anhui Province. The scenic spot is located in the northern part of the evergreen broadleafed forest in eastern China, at the intersection of subtropical and temperate zones.

There are 10 ecosystem types and 16 main vegetation groups in the Huangshan Scenic Spot. The landform types here can be divided according to changing altitude from high to low as: the gentle hilltop area, the intensive granite cutting zone, and the low, steep mountain area around the foothills. The special geological structure and topography have led to a slow process of vegetation restoration and evolution in the Huangshan Mountains. Once the ecosystem is damaged, it is thus very difficult to recover.

2.2.2 Special Habitat and Dense Distribution of Communities

The long-term orogeny, crustal uplift, and glacial and natural weathering have contributed to the formation of a unique peak forest structure in Mount Huangshan. In this structure, the massive peak forest is represented by Cinnabar, Eyebrows and Turtle, while the bamboo-slip peak forest is represented by Lotus, Tiandu, and Lotus Pistil peaks. From the Lion Forest to the Furong Mountains on the northern slope, there are several ridges arranged in parallel running north and south. These ridges are thin as blades and are thus called the knife-edged ridges. Furthermore, on both sides of the Paiyunting deep valley in the West China Sea, there are many steep and broken peaks.

The vertical peak forest is the most distinctive special ecological environment in Mount Huangshan, and the scarce soil makes it difficult for plants to attach and grow. The main plants growing in the vertical zone are *Pinus taiwanensis* and *Dendrocalamus chinensis*, and the peculiar landscape of Mount Huangshan is formed by both pine and strange stone formations.

2.2.3 Rich Array of Rare and Endangered Species

The Huangshan scenic area is the concentrated distribution area for the endemic species in the region, and there are more than 31 endemic plants with the word "Huangshan" in their names, such as *Rhododendron anhweiense*, *Kirengeshoma palmata*, *Sorbus amabilis*, and *Magnolia cylindrica*. At the same time, there are 21 species of wildlife under national protection in the scenic area, including 3 species of first-class protected plants and 18 species of second-class protected plants. In addition, there are 72 species listed in the *Red List of Chinese Species*; 26 species of protected plants listed in Appendix II to the *Convention on International Trade in Endangered Species of Wild Fauna and Flora* (CITES), all belonging to Orchidaceae; 136 ancient and famous trees listed under national key protection, 54 of which are listed in the *World Heritage Protection List*; 26 species of national protected animals in Mount Huangshan, of which 7 are under first-class protection and 19 are under second-class protection. Facing a fragile ecosystem and the rapid development of the tourism industry in scenic spots, these rare and unique plant species are in urgent need of better protection.

2.2.4 Various and Unique Genetic Resources

The Huangshan Scenic Area has a vast territory, superior ecological environment, abundant resources in its biological species, and contains a great amount of precious genetic diversity. In the surrounding areas of Huangshan, six leading industries and characteristic products have been established, namely the tea industry, the bamboo and timber industries, the silk industry, the fruit and vegetable industry, the traditional Chinese medicine industry, and the aquaculture industry.

There are many rare and endangered medicinal plant resources in Huangshan, which have important economic, scientific research, and medicinal value. For example, the wild soybean here is of great significance in cultivating or improving new varieties, and *Monascus fortunei*, *Liriodendron chinensis*, and *Prunus mume* are of great significance to the phylogenetic and taxonomic aspects of the subflora and flora. Thanks to its unique ecological environment, Mount Huangshan retains many ancient and rare tree species from the Quaternary glacial period, such as *Taxus wallichiana*, *Cercidiphyllum japonicum*, and many others, making it a unique green treasure house in China.

2.3 Cultural Value of Biodiversity in Huangshan Scenic Area

2.3.1 Wide and Profound Huizhou Cultural System

Located in ancient Huizhou, Huangshan has formed a Huizhou cultural system based on the core concepts of "harmony", "goodness", and "Confucianism", and the connotations of the conceptual, systemic, and local culture have developed over the course of its roughly 800 years of evolution, which has distinct regional characteristics. At the same time, Huangshan has Huizhou culture at its core, and the physical culture is based on the ancient buildings and villages, while the non-physical culture is derived from on the folk art and folk culture, the cultural schools of Xin'an Neo, Huizhou New Theory, the Xin'an School of Painting, Xin'an Medicine, and the Huizhou Confucian merchant culture.

2.3.2 Valuable Traditional Customs and Techniques

In Huizhou culture, the folk customs and techniques, including the marriage customs, temple fairs on the ninth day of the first lunar month, Zhongkui costumes in Yuliang, human pyramids, embroidered ball throwing, Nuo dance, and Mulian Drama all show unique local characteristics. Moreover, the traditional folk arts, including Huizhou opera, four carvings, seal carving, and printmaking, are all of great historical and artistic value.

2.3.3 Rich Connotations of the Literary, Artistic, and Religious Treasures

In Huangshan, more than 20,000 poems and lyrics have been composed praising the mountain. Since the late Ming and early Qing Dynasties, the "Huangshan Painting School" has created numerous works of calligraphy, painting, and photography themed on Mount Huangshan. The scenic spots of Mount Huangshan include Ciguang Pavilion, Banshan Temple, Yungu Temple, Songgu Temple, Cuiwei Temple, and other historical and cultural relics from Taoism and Buddhism, along with nearly 100 ancient structures, including ancient roads, bridges, temples, and pavilions. There are more than 200 stone carvings on the cliffs, which are an indispensable part of the humanistic landscape.

2.3.4　The Characteristic Architectural Style of the Residential Buildings

The folk houses, ancestral halls, and archways in the Huangshan area share the ancient Huizhou characteristics, and are also known as the "Three Great Ancient Buildings". Among the more than 5,000 surface monuments discovered, 2 ancient villages, Xidi and Hongcun of Yixian County, have been listed beside Mount Huangshan as World Cultural Heritage Sites. Furthermore, Mount Huangshan has 61 key national and provincial protection units.

2.3.5　The Interaction of Nature and Humanity

The Huizhou culture and Huangshan climate resources complement each other, with the natural environment and human environment blend together, forming a regional culture with Huangshan characteristics. Huangshan tea and the Huizhou tea ceremony each complement the other. The four treasures of Huizhou study are famous both at home and abroad for their raw materials and traditional craftsmanship. There are also more than 800 kinds of edible fungi and bamboo shoots, 80 kinds of mammals, and diverse freshwater fishes, which have provided an excellent material basis for the formation and development of Huizhou cuisine.

2.4　Evaluating the Value of the Biodiversity in the Huangshan Scenic Spot

2.4.1　Value Evaluation of the Direct Development of Biodiversity

Direct development value refers to the most intuitive market economy value produced by the development and utilization of the biodiversity and ecological assets of the Huangshan Scenic Spot, which mainly includes the economic benefits of leisure and recreation, scientific research, cultural, and educational activities.[①]

The direct development value of the Huangshan Scenic Spot is divided into four parts: tourism income value, scientific research value, cultural value, and product value. According to a preliminary estimation, the direct development value of Huangshan Scenic Spot is about 35.12×10^8 yuan per year, and the proportion of the direct value generated

---

① Ding Hui, Xu Haigen. Value Assessment of the Conservation and Utilization of Biological Species Resources: A Case Study of Jiangsu Province [J]. Journal of Ecology and Rural Environment, 2010, 26 (5): 454-460.

by tourism income is more than 80%. This data is shown in Figure 2-3.

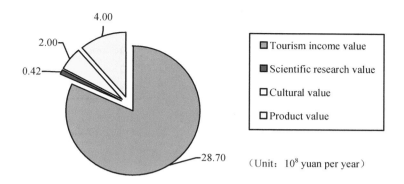

Figure 2-3　Estimation of the Direct Development Value of the Huangshan Scenic Spot and its Composition Diagram

2.4.2　Value Evaluation of the Ecological Protection of Biodiversity

Eco-protection value refers to the natural support for the normal production and consumption of tourism in the Huangshan Scenic Spot, that is not directly reflected by monetary value. It also provides adjustment and support services for the regional ecological environment, including ecological benefits and biological resources. According to a preliminary estimation, the ecological protection value of the Huangshan Scenic Area is about 208.54×10^8 yuan per year, in which the value of biological resources accounts for a dominant proportion (about 95%).

(1) Value of the Ecological Benefit

Ecological benefit value mainly refers to the functions of the natural ecosystem of the Huangshan scenic area in terms of water conservation, soil conservation, climate regulation, natural disaster control, and environmental purification. By measuring the monetary value of each part of the function according to the formula and accounting method, it has been concluded that the ecological benefit value of the Huangshan Scenic Spot is about 1.116×10^9 yuan per year. The functional composition and estimation results are shown in Figure 2-4.

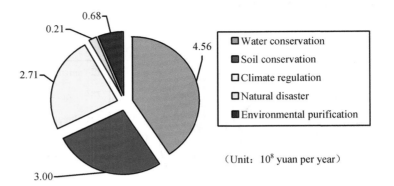

Figure 2-4 Estimation chart for the ecological benefit value of the Huangshan scenic area

①Water Conservation Function

This function mainly includes the infiltration and storage capacity of the ecosystem for water resources. Its value is mainly reflected in increasing the effective water quantity, improving the water quality, and regulating runoff. It is usually estimated by the market pricing method, that is:

Value of the conserved water source (yuan/year) = annual total amount of the conserved water source (t) $\times$ current residential water price (yuan/t)

Note: It is usually assumed that the total amount of conserved water sources per year = annual runoff = annual average precipitation - annual average transpiration

According to the hydrological and water-related data for the Huangshan Scenic Area, the average annual total amount of water conservation in the Huangshan Scenic Area is about 233.67×10^6 t, and the price of agricultural water is about 1.95 yuan/t. Therefore, the function has estimated the annual value for water conservation in the Huangshan Scenic Area to be about 4.56×10^8 yuan/year.

②Soil Conservation Function

Soil is an important non-renewable resource, and it is accumulated via the long-term biological and physical processes of the natural ecosystem. Its value lies in providing a place for the survival and development of organisms, degrading organic matter, maintaining fertility to ensure nutrient supply, and promoting the natural circulation of the earth. Therefore, it is an important aspect of the national and regional wealth.

The opportunity cost method can be used to evaluate the value of soil erosion prevention. The calculation formulas are as follows:

Total amount of soil erosion reduction = difference of soil erosion per unit area × total area of the ecosystem

Waste land area = total amount of soil loss reduction/average surface soil thickness

Value of soil erosion reduction = average benefit of ecosystem production × equivalent area of abandoned land

In addition, the value of preventing soil nutrient loss is generally measured by the total content of nitrogen, phosphorus, and potassium, which can be calculated by the market pricing method. The formula can thus be considered as follows:

Value of reducing soil nutrient loss = total soil loss × bulk density of soil layer × unit quantity of nitrogen, phosphorus, and potassium in the soil × market price of these three nutrients

According to statistical data, the total amount of soil erosion reduction in Huangshan City has reached 250 km^2, equivalent to 72,800 tons of chemical fertilizer, and the value of reducing soil nutrient loss has reached about 3.00×10^8 yuan per year.

③Climate Regulation Function

The production processes of green plants in the ecosystem can fix the carbon dioxide and regulate the oxygen exchange in the atmosphere, thus ensuring the basic climatic conditions of life activities. Therefore, climate regulation can be preliminarily estimated by the carbon sequestration and oxygen release method. The formulas are as follows:

Carbon Fixation-Carbon Tax Method: Carbon Fixation Value = Ecosystem CO_2 Absorption × Shadow Price

Oxygen Release-Alternative Cost Method: Oxygen Release Value = Total Biomass × 1.2 × Industrial Oxygen Production Unit Cost

The coefficients for carbon and oxygen are determined by photosynthesis.

Fixed Carbon = Fixed Carbon Dioxide × 0.27

Oxygen release = plant yield × 1.2

According to the preliminary statistics, the carbon storage of forest vegetation in the Huangshan scenic area is about 57,742.55 t, equivalent to 213,861.296 t of CO_2. The value of carbon storage in Huangshan scenic spot is 2.71×10^8 yuan per year, based on the shadow price of 19.11 yuan/t in the carbon emission trading market in Hubei Province in 2017.

④Disaster Control Function

The disaster control function mainly includes the functions of preventing natural and biological disasters such as droughts, floods, and landslides. The cost aversion method can be used for estimating, and the formula is as follows:

Value of natural disaster prevention = total ecosystem area × loss per unit area
Value of controlling biological disasters = input cost per unit area to prevent biological disasters × ecosystem area

According to the technical documents issued by the State Forestry Administration, the annual benefit to forest land for flood and drought mitigation is about 65 yuan/hm², and the forest land area of Huangshan Scenic Area is 12,271.1 hectares. The ecological benefit of flood and drought mitigation in the scenic area is 797,621.5 yuan/year. According to the statistics, the investment in the Huangshan Scenic Area for forest fire prevention, forest disease, and insect pest control is about 2×10^7 yuan per year. Therefore, the value of the disaster control function is about 2.1×10^7 yuan per year.

⑤Environmental Purification Function

This function estimates the value of purifying the ecosystem through the material ecological cycle. The alternative cost method can be used to estimate the cost. The formula is as follows:

The value of waste disposal = total ecosystem area × unit quantity of pollutants absorbed by the ecosystem × unit cost of industrial purification of pollutants

The absorption and purification functions of dust and harmful gases for forest land in the Huangshan Scenic Area were preliminarily generated. According to technical documents from the Forestry Bureau, the dust retention capacity of broad-leaved forest is 10.11 t/ (hm²·a), and the cost of reduction is 170 yuan/t. The amount of SO_2 absorbed is 88.65 kg/ (hm²·a), the cost of reducing SO_2 is 600 yuan/t; the amount of NO_x absorbed is 0.38 t/ (hm²·a), and the cost of reducing NO_x is 250 yuan/t. According to the preliminary estimation for the forest area of the Huangshan scenic area (12,000 hm²), the service value of purified gas in the Huangshan scenic area is about 2.3×10^7 yuan per year.

In addition, the value equivalent of Xie Gaodi's ecological service function was used to provide a preliminary estimate of the value of the waste disposal function for the

Huangshan Scenic Area, at about 4.5×10^7 yuan per year. The total value of the environmental purification function for scenic spots is about 6.8×10^7 yuan per year, as shown in Figure 2-5.

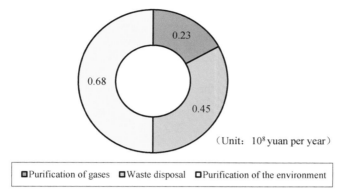

Figure 2-5 Value estimation and composition sketch of the environmental purification function for the Huangshan Scenic Area

(2) Value of Biological Resources

Biological resources are important for realizing a sustainable development strategy, and a people's practices for using biological resources results from their theoretical research. Therefore, understanding and studying the value of biological resources is of great importance in the decision-making for regional economic development strategies and species protection. Biological resources themselves are valuable. Some of them can enter the market directly and contribute economic value, thus their value can be clearly marked through their economic benefits. Others provide continuous services for human beings, but it is difficult to directly mark their value. At present, there are many methods to determine the value of biological resources in the world, which vary according to the standards and conditions of different countries. There are three evaluation perspectives that have been recognized:

①Evaluating the biological value in the natural state.

②Evaluating the value of biological products with commercial benefits.

③Evaluating the value of biodiversity services in ecosystem services.

The value of biological resources in the Huangshan Scenic Area is composed of three parts (Figure 2-6). Since there is no clear method for estimating the value of biological resources, only the representative biological resources with characteristics in Huangshan area are estimated. The total value is about 1.9738×10^{10} yuan per year.

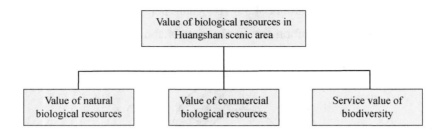

Figure 2-6　A sketch map for the value composition of biological resources in the Huangshan Scenic Area

①The Value of Biological Resources under Natural Conditions

Natural biological resources play an extremely important role in the process of regional development and in the production and life of residents. They are also important cultural relics and tourism resources for scenic spots. Among these resources, the value of food production and raw materials supply is preliminarily estimated based on the eco-service value equivalent factor table developed by Xie Gaodi et al., as shown in Table 2-5.

Table 2-5　Estimation of the Biological Resources Value in the Natural State of the Huangshan Scenic Spot

Category	Woodland	Grassland	Farmland	Wetland	Water Body	Wasteland	Town
Food Production Equivalent	0.1	0.3	1.0	0.3	0.1	0.01	0.00
Unit Area Value of Food Production（yuan/hectare）	201.22	603.65	2,012.15	603.65	201.22	20.12	0.00
Value of Food Production（10,000 yuan/year）	303.17	0	10.66	0	2.56	0	0
Equivalent Supply of Raw Materials	2.6	0.05	0.1	0.07	0.01	0.00	0.00
Unit Area Value of Raw Material Provision（yuan/hectare）	5,231.6	100.61	201.215	140.85	20.12	0	0
Value of Raw Material Supply（10,000 yuan/year）	7,882.45	0	1.07	0	0.26	0	0
Value of Natural Biological Resources（100 million yuan/year）	0.82						

In addition to providing biological resources for food production and raw materials, there are many unique and representative natural biological resources in the Huangshan area. The existence of these resources has also become a unique symbol of the Huangshan Scenic Area in the same industry in China, and refers to several types of resources that provide a reference for value assessment:

a. Value of Ancient and Famous Trees: Taking *Pinus taiwanensis*, *Mei taiwanensis*, *Ginkgo biloba*, and other famous ancient trees as representative varieties, according to *Huizhou Ancient Trees*, there are about 110 kinds of ancient trees in Huangshan, the earliest of which can be traced back to the Tang Dynasty. The "strangely-shaped pines" of the "Four Wonders" in Huangshan are all *Pinus taiwanensis*, and they are also a prime destination for tourists when enjoying the scenery of Mount Huangshan.

Based on the evaluation of the resource value of the ancient and famous trees in Mount Jiuhua from the available research data, the resource value of ancient and famous trees in Mount Huangshan scenic area is estimated by referring to its primary data, which is shown in Table 2-6.

Table 2-6 Value Estimation of the Ancient and Famous Tree Resources in the Huangshan Scenic Area

	Visitor Reception (10,000 people)	Amount 1 willing to pay (100 million yuan per year)	Amount 2 willing to pay (100 million yuan per year)	Total Mean Value (100 million yuan per year)
Mount Jiuhua Scenic Area	401	1.500	4.589	3.04
Mount Huangshan Scenic Area	299.1	1.119	3.423	2.27

b. Value of Rare Animals and Plants: In the Huangshan scenic area, there are three species of first-class protected plants, 18 species of second-class protected plants, 7 species of first-class protected animals, including 2 species of birds, and dozens of species of endemic plants and animals, such as Macaca thibetana huangshanensis, *Huperzia Huangshan*, and *Deutzia glauca Cheng*.

② Value of Commercial Biological Resources

a. Medicinal Biological Resources: There are many kinds of medicinal herbs in Huangshan, and their reserves are quite different. According to the census statistics for traditional Chinese medicine resources, there are more than 400 kinds of commonly used bulk medicines. Among them, 66 species, such as *Peucedanum praeruptorum*, *Stephania tetrandra*, *Smilax glabra*, *Albizia julibrissin*, *Sargentodoxa cuneata*, *Florists Chrysanthemum*, *Sedum sarmentosum*, *Hawthorn*, *Rubus chingii* and *Bamboo Juice*, have an annual yield of more than 100 tons. 131 species, such as (Hui) *Atractylodes macrocephala*, *Fallopia multiflora*, *Morus alba*, *Magnolia denudata*, *Armeniaca mume*, *Lysimachia christinae*, *Cornus officinalis*, *Fructus Ligustri*, *coix seed*, *Lygodium japonicum*, *Rhus chinensis* and *bat guano*, have an annual yield of 10 to 100 tons. Based on their market prices, the total value of these medicinal resources is estimated to be 1.38 billion yuan per year.

b. Tea Resources in Huangshan: According to the statistical data for Huangshan, the output of Huangshan spring tea in 2013 was about 12,966 tons, which was basically the same as the data for the same period last year. The average price of Huangshan spring tea was 61.8 yuan/kg, an increase of 25.1% over the same period last year, and its output value reached 801.65 million yuan.

In addition, according to the evaluation information of Chinese brand value released by the CCTV Financial Channel and the China Brand Construction Promotion Association in 2015,Huangshan Taiping Houkui, a regional tea brand (protection of geographical indications), has a brand strength of 920.0 and a location brand value of 11.143 billion yuan. In addition to Taiping Houkui, famous teas like Huangshan Maofeng and Keemun Black Tea are also internationally well-known brands. Based on this, the brand value of Huangshan Tea is estimated to be about 15 billion yuan.

c. Wild Mountain Delicacy Resources: Hui cuisine is one of the eight major cuisines in China, and the mountain delicacies are the main feature of Hui cuisine. Phyllostachys pubescens shoots, wild mountain shoots, ferns, and edible fungi are the main local specialties in the scenic spots.

d. Traditional Cultural Resources in the Region Based on Biological Materials: The Huangshan scenic area has handicrafts, Chinese medicinal materials, and other geographically distinct products. For example Xuan Paper's brand strength has reached 920.0,and its regional brand value reached 2.149 billion yuan.

③ Service Value of Biodiversity

The Huangshan scenic area is rich in animal and plant resources and has great potential value. This part of the value can be estimated by the opportunity cost method or the replacement cost method. It can also provide a preliminary estimate of the ecological value of biodiversity conservation based on the eco-service value equivalent factor table developed by Xie Gaodi and other research, the results of which are shown in Table 2-7.

Table 2-7　Estimation Table for the Biodiversity Value of the Huangshan Scenic Area

	Woodland	Grassland	Farmland	Wetland	Water Body	Wasteland	Town
Value Equivalent of Biodiversity	3.26	1.09	0.71	2.50	2.49	0.34	0.00
The Unit Area Value of Mount Huangshan (yuan/hectare)	6,559.62	2,193.25	1,428.63	5,030.38	5,010.26	684.13	0.00
The Value of Biodiversity of Various Types in Huangshan Mountains (10,000 yuan/year)	9,883.38	0.00	7.57	0.00	63.63	0.00	0.00
Biodiversity Value of Mount Huangshan (100 million yuan/year)	1.00						

2.4.3　The Value of the Future Existence of Biodiversity

The value of future existence refers to the value of Huangshan which is not directly exploited and utilized at present, but which objectively exists as Huangshan resources and can be used by future generations or the people themselves, including the value of existence, heritage, and choice. Among these, the value of existence refers to the fees that people pay voluntarily to ensure the sustainable existence of Huangshan tourism resources. Heritage value refers to the fees paid by contemporaries to retain Huangshan Mountain as a tourist resource for future generations. The value of choice is an insurance premium that individuals voluntarily pay in advance each year, so that their descendants or others can selectively use Huangshan tourism resources in the future. For these types of ecosystem services, only a hypothetical market evaluation method, the willingness survey method (CVM), can be used.

By estimating based on the research literature, statistical data, and economic growth, the maximum per capita willingness to pay for the protection of the Huangshan ecosystem in 2017 is about 27.96 yuan, while the total population of China in 2017 is 1.39 billion, and the permanent population in cities and towns is 813.47 million. Based on this, the future value of the Huangshan Scenic Area is preliminarily estimated to be from 227.45×10^8 yuan per year to 388.67×10^8 yuan per year.

2.4.4　Value Evaluation of the Biodiversity in the Huangshan Scenic Spot

The composition and estimation results of the biodiversity value for the Huangshan Scenic Area are shown in Figure 2-7. Among these results, the value of ecological protection and the value of future existence are categorized as indirect use value, which cannot be reflected by currency, but their estimated value is much higher than the income of direct development, which means that the Huangshan scenic area is well worth protecting.

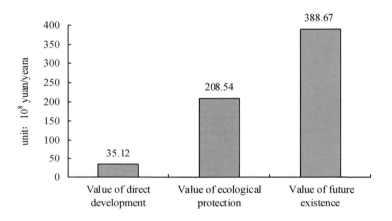

Figure 2-7　An Outline of the Economic Value Composition of the Biodiversity in the Huangshan Scenic Area

From the point of view of sustainable development, the direct development value is the market economy benefit that the contemporary people obtain by utilizing Huangshan tourism resources. The ecological protection value embodies the ecological benefit of the Huangshan Scenic Area for the regional ecological environment and ecological security. The value of future existence reflects the benefits that future generations can derive from the existence of Mount Huangshan. In terms of value, the protection of Mount Huangshan should be more important than its development. Protection thus comes first and development second.

The great use value and existence value of Mount Huangshan are closely related to its unique geological landform, diverse biological resources, beautiful landscape and profound cultural heritage. But these unique landscapes are easily destroyed by the interference of unreasonable human activities, which thus reduces the value of the tourism resources of Mount Huangshan.

Through protection, the tourism resources of Huangshan can be sustained, and the value of development, ecology, and existence can all be enhanced. If Mount Huangshan is well protected, its development will be supported by its resources. The economic benefits generated by development are embodied by the use value. The better the protection, the value that can be realized through development will be accordingly improved. At the same time, through development, we can raise funds for the protection of Huangshan Mountain, promote its protection, and produce higher economic value in the future.

2.5 Conservation and Utilization of the Biodiversity in the Huangshan Scenic Area

2.5.1 Conservation and Management of the Biodiversity in Scenic Spots

Initiating the system of "alternate rest" in domestic scenic spots. In October 1987, Huangshan initiated a system of "alternate rest" for scenic spots in China. During the period of closure, the natural ecosystems and wildlife resources of the scenic spots are restored both naturally and artificially. Each alternating period lasts from 3 to 5 years. Classical scenic spots such as Lianhua Peak, Lion Peak, Danxia Peak, Tiandu Peak and Shixin Peak have all been successively closed. Thus, the natural vegetation types and vertical vegetation zones in the transitional zone from Central Asia to North Subtropics are effectively maintained.

Strengthening the protection of ancient and famous trees. There are 136 ancient and famous trees registered and archived in the Huangshan Scenic Area, of which 54 are listed in the World Natural Heritage List. The scenic area encloses the trees on both sides of the tourist road with pieces of bamboo, carries out regular comprehensive investigation and prevention of damage to the ancient and famous trees, invites experts to carry out physical examinations and consultation, takes comprehensive protection measures, especially employs staff for daily maintenance, as well as compiling management software and carrying out information management for the ancient and famous trees.

Strictly carrying out plant quarantines and fire control measures. A monitoring network of diseases and insect pests focusing on prevention of pine wood nematode disease has been established, and monitoring stations, sub-stations, monitoring points, and quarantine checkpoints have been set up in the pine forests in the scenic area. A special survey for pine wood nematode disease is carried out in the spring and autumn every year, and dead trees are cleaned up in a timely manner. The scenic spot attaches great importance to the construction of its fire prevention infrastructure, fire prevention propaganda, fire source monitoring, and cooperative joint defense, achieving excellent results and 39 consecutive years without forest fires.

Actively developing water resource protection. Ponds, reservoirs and other water storage facilities have been successively built to meet the water demand, strictly control the discharge and end-treatment of wastewater and waste in scenic spots, limit the number of hotels and restaurants, protect the water quality and water sources from the source, provide a good habitat for the many species, and benefit the protection of biodiversity. In addition, the high water quality of the Qiantang River Basin and Jiangsu and Zhejiang Plains has been ensured, and the water safety of residents in the basin has been guaranteed.

2.5.2 Utilization of Biodiversity in Scenic Spots

An excellent natural ecological endowment and outstanding tourism development benefits. The natural vegetation of the Huangshan Scenic Area is well preserved, with a high degree of aggregation and good connectivity of the patches, which is conducive to providing a complete habitat for endangered species. The rock landscape has a high degree of fragmentation, which provides a variety of stone forest landscape for tourists. In addition, the region is well known for the "five unique features", namely strangely-shaped pines, rocks with intriguing shapes, a sea of clouds, hot springs, and winter snow, which have all been developed. Based on these spectacular sights, the hot spring scenic spot, the Yungu scenic spot, the Yuping scenic spot, the Beihai scenic spot, the Tianhai scenic spot, the Xihai scenic spot, the Diaoqiao scenic spot, and the Donghai scenic spot have all been developed and established. Since opening in 1979, the number of tourists and the number of overseas visitors has been ever increasing, and the tourism market has become increasingly mature.

The high value of biological resource development accelerates the development of surrounding communities. Biological resources such as the forests, Chinese medicinal materials, flowers, and crops in Mount Huangshan have great potential for development. Apart from the tourism service industries of the residents in the five towns and the woodland around the scenic area, there is under-forest cultivation and aquaculture, the planting and marketing of traditional Chinese medicines, original tea production, flowering bonsai production, and other characteristic industries that have flourished and become an important means for residents to increase their income.

The development of the humanistic scenic areas is accelerating and interacting with the utilization of natural resources. The humanistic scenic areas and artistic works are well preserved in the Huangshan scenic area, and the literature, calligraphy, painting, photography, and other works are endlessly emerging. Numerous natural and humanistic scenic spots, such as Tunxi Old Street, Mount Qiyun in Xiuning, Xidi and Hongcun in Yixian county, the Tangyue memorial archway in Shexian County, and the ancient city of Huizhou, have initially formed the comprehensive ecological products of the Huangshan scenic area. In addition, there is great potential for the development of recreational and healthy lifestyle vacations, such as forest walks, hiking, pastoral health care, hot spring convalescence, gourmet tea products, and many others. The cultural system in Huizhou, with its distinctive regional characteristics, intersects with the climatic resources of Huangshan Mountain, and the natural and cultural environment integrate with each other, forming the regional culture, which has crystallized in Huangshan tea, the Huizhou scholar's four treasures, Xin'an medicine, the Xin'an School of Painting, and Huizhou cuisine, all of which highlight the unique characteristics of Mount Huangshan.

3 Evaluation of the Environmental Capacity and Bearing Capacity for Tourism in the Huangshan Scenic Area

3.1 The Restriction of Tourism Development and Environmental Bearing Capacity in the Huangshan Scenic Area

3.1.1 Characteristics of the Development of the Tourism Industry in the Huangshan Scenic Area

(1) There Will Be a Period of Explosive and Sustained Growth

As far as the country is concerned, with the increase in income of the residents and the gradual implementation of "paid vacation" and the arrival of the automobile era, people's spending on tourism is continually increasing. Tourism will thus become a rigid demand. The form of tourism will change from sightseeing tourism to leisure tourism and vacation tourism, and the era of mass tourism will arrive in a comprehensive way. From a regional perspective, the Huangshan Scenic Area has a superior geographical position, and people in the Yangtze River Delta and in the middle reaches of the Yangtze River can even drive to Huangshan themselves. Moreover, the addition of flights and high-speed trains has brought more visitors to Huangshan from the both whole country and even the whole world. Therefore, a growth trend in the number of visitors can be expected.

(2) "Sightseeing Tourism" and "Short and Medium Distance Tourism"

At present, the Huangshan scenic spot still holds sightseeing as its main project, while recreational and leisure cultural tourism projects remain relatively few. In terms of the regionality and time spent for tourism, short-distance tourism is the main form, while long-distance tourism is relatively scarce. Visiting Huangshan does not take much time, usually two or three days or three or five days, with "one-day tours" accounting for a large proportion, while long-term tourism remains relatively rare. Most of the leisure health care and vacation tourism are mainly for the elderly from the surrounding cities, and it still only occupies a small proportion.

(3) Intelligent Scenic Spots and Big Data Management

The Huangshan Scenic Spot will accelerate the construction of intelligent systems to enhance its competitiveness in the tourism industry. Currently, an information network covering the whole mountain has been established, which has played an active role in the protection of scenic resources, scenic area management, tourism services, business development, and other aspects. The "Huangshan Tour with a QR code"Intelligent Tourism Service Platform, passenger flow forecast, and Baidou navigation have accelerated the development of the scenic spots.

3.1.2 Main Pressure on the Environmental Bearing Capacity of the Huangshan Scenic Spot

(1) Pressure on the Spatial Bearing Capacity of Resources

This problem is particularly evident in some scenic spots and corridors, such as the upper and lower stations of the ropeway, in front of the Beihai Hotel, Bright Summit, Crucian Carp Back, the Dream of Being a Successful Writer viewing platform, and other areas. The problem of resource space overload in Mount Huangshan has long attracted the attention of the managers, who have hence taken measures, but the problem has still not been thoroughly solved.

(2) Pressure on the Ecological Environment Carrying Capacity

Tourist activities have exerted some degree of subtle influence on the ecological environment of the Huangshan Scenic Spot, which have gradually emerged with the passage of time. Taking the "strangely-shaped pines" of the "Four Wonders" in Huangshan for example, there are often wounded or dead pine trees on both sides of scenic spots, or mountain paths that are frequently passed by crowds, which are directly related to human destruction during the peak season of tourism, human trampling that results in soil hardening, poor drainage, and damage to the tree roots.

(3) Pressure on the Water Supply

The water supply has always been one of the main environmental constraints in the development of tourism in Huangshan. Due to insufficient water resources and poor water storage conditions, tourists and workers often have difficulties using water. The scenic spots with the most serious water shortage are Beihai, Yuping, and Diaoqiao, and the largest water shortage period is from September to October. The insufficient water supply not only makes it difficult for tourists to use water, but also affects vegetation protection in the scenic spots, because the natural stream and spring water will be artificially intercepted and transported to hotels, which then causes a water shortage for vegetation and damages the aesthetics of the water landscape.

(4) Pressure on the Psychological Bearing of Tourists

During the peak season for tourism, tourists will wander all over the mountains and you even need to step aside from the crowds to see the scenery, which is a common problem that occurs in all scenic spots throughout the country. Especially during the National Day "Golden Week for Tourism", and the short and long holidays, due to the problems with queuing for the ropeway, toilet congestion, and the overcrowding of scenic platform during the peak season of Huangshan tourism, the psychological carrying capacity of tourists is usually overly stressed.

3.2 Estimation of the Environmental Capacity for Tourism in the Huangshan Scenic Area

3.2.1 Space Environmental Capacity

Through the method of instantaneously calculating the space of scenic spots, we have estimated the maximum number of tourists that the Huangshan scenic spots can bear in the half a year of peak season in the summer (12 hours per day). The main space environment in the scenic spots can be divided into two parts, namely, the line sightseeing area and the platform sightseeing area, which can be calculated by the line method and area method respectively. According to the calculation standard for tourist capacity in the *Standard for Scenic Spot Planning* (GB 50298-1999) (Table 3-1), we have calculated the daily turnover rate, with the average time for tourists completing the main scenic spots in the whole scenic area, and the sum up the spatial environmental capacity for the main scenic spots.

Table 3-1 Calculation Criteria for the Space Environmental Capacity of the Huangshan Scenic Spot

Primary Category	Basis for Capacity Measurement	Measuring Criteria
Sightseeing Path for Motor Vehicles	According to the number of motor vehicles, speed, and road width of the tourists in the evacuation scenic area, a standard tourist car with 25 passengers can be set up every 1-2 kilometers on a single lane.	25 m/person
Pedestrian Sightseeing Path and Platform	According to the *Regulation of Scenic Spot Planning*(GB 50298—1999), the standard value of capacity calculation is 5-8 m^2/person, the limit value is 5 m^2/person, and the reasonable capacity of a non-peak day is 8 m^2/person.	5-8 m^2/person
Classical Scenic Spots	At the small tourist attractions, according to the psychological capacity of tourists, the standard value is 1-1.5 m^2/person, and the limit value is calculated according to 1 m^2/person, with the reasonable capacity of a non-peak day calculated at 1.5 m^2/person.	1-1.5 m^2/person
Ecotourism Routes	In order to meet the needs of tourists for field exploration, some scenic spots have added new routes. The average width of the road is 1 m and the capacity is 5 times the standard value of the sightseeing track.	25-40 m^2/person

We have calculated the capacity of various types of space environment based on data from the *Huangshan Scenic Spot Master Plan* (2007-2025), the Huangshan Scenic Spot Tourist Guide Map (Figure 3-1), Diaoqiao Scenic Spot (Expansion) Planning (2007-2020), *Donghai Scenic Spot (Fugu Management Zone) Detailed Planning of the Huangshan Scenic Spot*, and construction of the lower half of the Yungu Cableway (design scheme of railway trail from the upper station to Shisungang Footpath) in the Huangshan Scenic Spot.

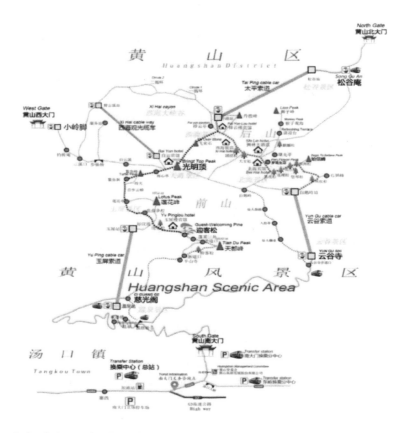

Figure 3-1　Schematic Map of the Main Tourist Routes in the Huangshan Scenic Area

Note：The pictures are from the official website of the scenic area. http://hsgwh.huangshan.gov.cn/News/show/1043916.html.

(1) Measurement and Calculation of the Environmental Space Capacity of the Pedestrian Walkways

According to the statistics, the length of the main sightseeing walkways in the Huangshan scenic area is about 74 km (with the length of new walkways being about 14 km), the length of newly established tour routes is about 18 km, the length of a tour is 12 hours in the half year around summer, and the limit value of the tourist capacity is 5 m^2/person (25 m^2/person on ecological tour routes). During the planning period (2018-2030), it is estimated that the maximum capacity of pedestrian paths in the scenic area in the half year around summer will be 53,763 people per day.

(2) Measurement and Calculation of the Environmental Space Capacity of the Motor Vehicle Sightseeing Road

The statistics have shown that the length of main motor vehicle sightseeing paths is about 29 km（with the length of new motor vehicle routes being about 16 km） during the planning period for the Huangshan Scenic Area（2018-2030）. According to a travel time of 12 hours in the half year around summer and the limit value of the tourist capacity of 25 m，it is estimated that the maximum tourist capacity for the motor vehicle route in the half year around summer is 3,146 persons per day.

（3）Measurement and Calculation of Environmental Space Capacity of Observation Platform and Scenic Spot

During the planning period of the Huangshan Scenic Area（2018-2030），the area of the main tourist platforms（including scenic spots）is about 150,000 m^2. According to the length of 12 hours in the half year around summer and the limit value of the tourist capacity of 5 m^2/person（1 m^2/person in the classical scenic spots），the maximum tourist capacity of the scenic platforms at the scenic spots in the half year around summer（per spot） is 7,936 people/day.

（4）Comparisons of the Spatial Environmental Capacity and the Entry and Exit Channels of the Scenic Spots

Presently，most tourists in the Huangshan Scenic Area choose Nanmen（Tangkou）as the main entry and exit channels. There are many sightseeing trucks，and the Hot Spring scenic area and Yungu scenic area have become the prime tourist destination for most tourists. In terms of spatial relations，the hot springs and Yungu scenic spots are entrances or signs for the Huangshan tourist area，where visitors have a fixed concentration space，short stay time，and high daily turnover rate. With the opening of the North Gate cableway and the Xihai sightseeing cable car，tourists have begun to be diverted to the North Gate and the West Gate. Therefore，the Diaoqiao scenic spot and the Xihai scenic spot will mainly play the role of diverting and extending tourists. Tourists stay there for a short time and can enjoy a larger area per capita，and thus the space capacity is also larger.

Yuping，Beihai，Tianhai，and Xihai are the core elements of the Huangshan scenic spot，and the tourist rates for Tiandu Peak，Lianhua Peak，Bright Summit，and Shixin Peak are nearly 100%. At the same time，in the Tianhai and Beihai scenic spots，there will be a large number of tourists needing catering，recreation，changing roads，and people gathering and distributing，therefore tourists will stay there for a longer time. These four scenic spots will thus become the entry and exit channels，affecting the tourist safety and the capacity of the Huangshan scenic area during the peak season.

Preliminary estimates show that during the planning period（2018-2030），the maximum daily capacity of the Yuping scenic spot in the Huangshan scenic area is 7,061,the maximum daily capacity of the Beihai scenic spot is 6,110,the maximum daily capacity of the Tianhai scenic spot is 9,579,and the maximum daily capacity of the Xihai scenic spot is 5,254. The daily limit capacity for visitors is 64,846,as shown in Table 3-2.

Table 3-2　Daily Environmental Space Capacity for Tourists in the Peak Season of the Huangshan Scenic Area　　　（Unit：person/day）

Scenic Spot / Capacity	Hot Spring Scenic Spot	Yungu Scenic Spot	Yuping Scenic Spot	Beihai Scenic Spot	Tianhai Scenic Spot	Xihai Scenic Spot	Diaoqiao Scenic Spot	Donghai Scenic Spot	Total2
Walkway Capacity	955	4,631	6,335	3,798	8,367	3,784	14,664	11,230	53,764
Vehicle Line Capacity	1,140	—	—	—	—	—	1,526	480	3,146
Platform Scenic Spot Capacity	—	664	726	2,312	1,212	1,470	1,112	440	7,936
Total 1	2,095	5,295	7,061	6,110	9,579	5,254	17,302	12,150	64,846

（5）Planned Capacity for Scenic Area Facilities

There are three cableways in the scenic area，at Yuping，Yungu，and Taiping，and four convenient routes for sightseeing cable cars in Xihai，and they mainly carry tourists up and down the hills. According to the estimate of the cableway capacity and turnaround time，the maximum daily cableway capacity is about 110,000 people，which can meet the overall needs of the tourists in scenic spots. Inevitably，in view of the instantaneous tourist capacity of the core scenic spots（as shown in Table 3-3），excessive instantaneous tourist volume will cause congestion and long wait times for the cableways during the peak season of tourism，as shown in Table 3-4.

Table 3-3　Instantaneous Tourist Capacity in the Core Scenic Spots of the Huangshan Scenic Area

	Yuping Scenic Spot	Beihai Scenic Spot	Tianhai Scenic Spot	Xihai Scenic Spot	Total 4
Instantaneous Capacity of Tourist Routes	2,371	1,597	4,183	1,468	9,619
Instantaneous Capacity of the Viewing Platform	190	862	534	450	2,036
Instantaneous Capacity of the Classical Scenic Spots	90	270	—	40	400
Total 3	2,651	2,729	4,717	1,958	12,055

Table 3-4 Estimation of the Capacity and Limit Waiting Time of Cableway Facilities in the Huangshan Scenic Area

Facilities \ Index	Maximum One-way Traffic (person/hour)	Running Time (hour)	Maximum Transport Capacity (person/day)	Instantaneous Tourist Extremes for a Single Trip	Expected Waiting Time
Yuping Cableway	2,400	10	48,000	12,055	5.0 h
Yungu Cableway	2,000	10	40,000	12,055	6.0 h
Taiping Cableway	600	10	12,000	12,055	—
Xihai Sightseeing Cable Car	800	10	16,000	12,055	—
Total	5,800		116,000		2.0 h

3.2.2 Eco-environmental Capacity

The carrying capacity of the ecological environment of a tourist destination depends on three variables: the ability of the natural ecosystem to purify and absorb pollutants, the ability of artificial systems to treat pollutants, and the number of pollutants produced per capita per unit time, including tourists, scenic management personnel, service personnel, and other permanent residents.

Compared with the natural purification capacity, the purification capacity of the artificial systems and the controllability of the amount of pollutants produced by tourists and the permanent population are stronger, and thus the calculation of the ecological environment carrying capacity refers more to the pollutant treatment capacity of the artificial system.

Table 3-5 Calculation of the Eco-environmental Capacity of the Huangshan Scenic Spot

Evaluation Projects	Formula	Parameters	Data Description	Capacity Estimation
Atmospheric Environmental Capacity	$B = \dfrac{S \times f}{S_m}$	$S = 160.6$ km^2 $f = 98.29\%$ $S_m = 1.7 \times 10^{-7}$ km^2/person	S_m was obtained through the per capita CO_2 emission and conversion rate. The per capita CO_2 emission was about 0.9 kg, and the average coefficient of CO_2 absorption in the forest land was 5.2 t/hm^2.	9.8×10^8

Evaluation Projects	Formula	Parameters	Data Description	Capacity Estimation
Environmental Capacity of Sewage Treatment	$B = \dfrac{X}{Y}$	$X=4.0\times10^6\ m^3/day$ $Y=0.082\ m^3/(person\cdot day)$	Sewage treatment volume is estimated at 166 tons/ hour of the planned value, and Y is the average value.	4.9×10^7
Environmental Capacity of Solid Waste Treatment	$B = \dfrac{W}{Z}$	$W=21\times10^3\ kg$ $Z=0.3\ kg/day$	Implementation of "unified standards, standardized operation, unified transportation, centralized disposal, unified pricing, cost sharing, unified production, sealed packaging" domestic waste move downhill in a comprehensive way	7.0×10^4

According to the calculation results shown in Table 3-5, it can be concluded that:

The daily ecological environment capacity $B=\min\{9.8\times10^8, 4.9\times10^7, 7\times10^4\}=70{,}000$ people

It should be added that the calculation of the ecological capacity from the garbage disposal capacity is based on the presumed daily production and clearance of garbage. In practice, the distribution of tourists is seasonally unbalanced. On the peak days of tourism, the garbage carriers and cableways can be increased. Therefore, the carrying capacity of solid waste can thus be appropriately higher than that calculated. If the time-varying coefficient is 2, the ecological environment capacity can reach 140,000 people per day.

3.2.3 Economic Environmental Capacity

To measure the economic and environmental capacity of the Huangshan Scenic Area, the accommodation, water supply, and transportation facilities must be considered. Among these, the accommodation facilities are divided into two parts: scenic area and outside the scenic area. In order to better protect the ecological resources of the scenic area, the number of hotels and restaurants inside the area is gradually being reduced, and more and more individual beds are being distributed in Tangkou, Zhaixi, and Shancha outside the scenic spot. In addition, with the continuous expansion of the Huangshan tour line, the hotels and residential accommodation on the Tunxi and Shexian ring lines also receive many tourists in the peak season. The rheological property of the accommodation facilities outside scenic spots is relatively large and usually plays a role in alleviating difficulties in finding accommodation in the peak season of tourism in these areas. Therefore, the environmental capacity of the accommodation facilities in scenic spots is not considered. We therefore only estimate the capacity of the water supply and transportation facilities in scenic spots.

(1) Capacity Measurement of the Water Supply Facilities in Scenic Spots

8 ponds with a total water storage capacity of 243,860 m^3, and 20 reservoirs with a total water storage capacity of 5,880 m^3 have been built in the Huangshan scenic area. In addition, there are hot spring water distribution stations with an annual water supply of 132,000 m^3, and the total water storage capacity of the water supply facilities is 381,740 m^3. According to the calculations for a dry year, the guarantee rate is 95%, the coefficient of reclamation is 3.5, and the utilization rate of the annual water supply is 90%. The amount of water the Huangshan scenic area provides every day is:

$$381,740 \times 95\% \times 3.5 \times 90\% \div 365 = 3,129.75 \ m^3/day$$

According to the *Code for the Planning of Scenic Spots* (GB 50298-1999), which was implemented on January 1, 2000, the water consumption per capita is as follows: the average water consumption for individual guests is 0.02 m^3/day and the average water consumption standard for hotels is 0.2 m^3/day for each bed. There are 5,367 beds in the scenic area, and the occupancy rate is 60%. The water consumption of tourists with hotel accommodations is 644 m^3/day. The upper limit of the daily water consumption for mountain workers is about 0.15 m^3/person/day $\times$ 865 persons = 129.75 m^3/day. The capacity of tourists provided by the water supply facilities is as follows:

$$(3,129.75 - 644 - 129.75) / 0.02 = 117,800 \ people$$

(2) Capacity Measurement of the Transportation Facilities in Scenic Spots

According to the statistical yearbook of the Huangshan Scenic Area in 2017, there are 155 tourist vehicles in the Huangshan scenic area, including 141 large buses, 10 medium buses, and 4 small buses. Assuming that the passenger carrying capacity of all types of vehicles is 55, 40, and 20, and that the shuttle buses are sent out every 20 minutes following the two main routes (the scenic area transfer center-Yungusi Temple and the scenic area transfer center-Ciguang Pavilion), the daily one-way transport capacity will be about 320,000 people, as shown in Table 3-6.

Table 3-6 Capacity Measurement of the Transportation Facilities in the Huangshan Scenic Area

Vehicle Types and Single Authorized Number			One-way Departure Frequency/h	Running Time (Summer)	Traffic Facility Capacity
Large Vehicle	141×50				
Medium-sized Vehicle	10×40	8,235	3	13 h	321,165
Light-Duty Vehicle	4×20				

3.2.4 Environment Capacity in Terms of Social Psychological

(1) Psychological Capacity of Local Residents

In the Huangshan Scenic Area, the residential area has been separated from the scenic area. The permanent population of the scenic area is the service personnel and management personnel directly engaged in tourism, and the psychological carrying capacity of these residents is usually close to infinite. Residents of the villages and towns around the scenic spot have participated in the tourism development of the Huangshan scenic area with labor force participation. Huangshan, especially the Tunxi District, where the municipal government is located, has always been the gateway of Huangshan's opening to the outside world and the gathering and distributing place for tourists, which has basically supported the optimistic attitude towards tourism development and the high acceptance rate of tourists. Therefore, at present and even in the future, the psychological carrying capacity of the local residents will not constitute a bottleneck for the tourism development in the Huangshan Scenic Area.

(2) The Psychological Capacity of Visitors

The psychological carrying capacity of tourists is based on their crowd sensitivity and varies according to their personality and behavioral characteristics. Psychological shortcomings in tourists often arise from scenic spots with high requirements for landscape viewing and high visibility, while tourists are more tolerant at general scenic spots. In the calculation, we only consider the psychological bearing capacity of tourists in the core scenic spots of Mount Huangshan. We take the psychological extreme value of 1.8 m^2/person for the tourist space, and calculate the capacity of the four scenic spots of Beihai, Tianhai, Xihai and Yuping. The psychological extreme value is 135,976.

3.2.5 Tourism Environmental Capacity

It can be inferred from the information above that the tourism environmental capacity of the Huangshan scenic area in the half year around summer under normal weather conditions is as follows:

$$TEBC_{sum} = \min\{64,846.7 \times 10^4, 117,800, 135,976\} = 64,846 \text{ people/day}$$

It can thus be seen that the spatial environmental capacity of the Huangshan scenic area is the corresponding value for the entry and exit passages of the tourist capacity of the scenic area. In winter half of the year, due to the shortening of travel times in each scenic spot (10 hours), the turnover rate of tourists changes (the speed of travel is slowed down). It is roughly estimated that the daily environmental capacity of tourism in winter half of the year is as follows:

$$TEBC_{win} = 48,660 \text{ people/day}$$

3.3 Carrying Capacity and Evaluation of the Tourism Environment in the Huangshan Scenic Area

3.3.1 Daily Carrying Capacity Analysis

(1) Reasonable Capacity for Tourists on Non-peak Days

According to the standard for a reasonable per capita occupied area stipulated in the *Standards of Scenic Spot Planning* (GB 50298-1999), the tourist capacity of each scenic spot on a non-peak day is estimated according to the average turnover rate of the tourist day. The result is shown in Table 3-7, and shows that the average reasonable tourist capacity of the Huangshan scenic area on a non-peak day is about 34,807.

Table 3-7 Estimation of the Reasonable Tourist Capacity on Non-peak Days in the Huangshan Scenic Area (Unit: person/day)

Scenic Spot \ Capacity	Hot Spring Scenic Spot	Yungu Scenic Spot	Yuping Scenic Spot	Beihai Scenic Spot	Tianhai Scenic Spot	Xihai Scenic Spot	Diaoqiao Scenic Spot	Donghai Scenic Spot	Total2
Walkway Capacity	271	1,464	2,694	1,815	4,754	1,669	8,332	5,784	26,783

Scenic Spot\Capacity	Hot Spring Scenic Spot	Yungu Scenic Spot	Yuping Scenic Spot	Beihai Scenic Spot	Tianhai Scenic Spot	Xihai Scenic Spot	Diaoqiao Scenic Spot	Donghai Scenic Spot	Total2
Vehicle Line Capacity	691	—	—	—	—	—	1,156	291	2,138
Platform Scenic Spot Capacity	—	302	509	2,058	971	785	1,011	250	5,886
Total 1	962	1,766	3,203	3,873	5,725	2,454	10,499	6,325	34,807

(2) Tourist Carrying Capacity and Regulation

According to the *National General Standards for the Management of Scenic Spots* (GB/T 3435-2017), the number of tourists is regulated based on the reasonable tourist capacity and limit tourist capacity. The maximum tourist capacity of a peak day and non-peak day in Huangshan are as follows:

$TEBC_{sum}$ = 64,846 people/day; $TEBC_{win}$ = 48,660 people/day; $TEBC_{avg}$ = 34,807 people/day

Table 3-8 Safety Early Warning Level for Visitors and Number in the Huangshan Scenic Area

Types of Early Warning	Early Warning Level	Early Warning Indicators	Population of Early Warnings
Peak Day of Visiting in Winter Half Year Extreme Tourist Capacity	Yellow Early Warning	$\geqslant$70% of $TEBC_{win}$	34,062
	Orange Early Warning	$\geqslant$80% of $TEBC_{win}$	38,928
	Red Early Warning	$\geqslant$90% of $TEBC_{win}$	43,794
Peak Day of Visiting in Summer Half Year Extreme Tourist Capacity	Yellow Early Warning	$\geqslant$70% of $TEBC_{sum}$	45,392
	Orange Early Warning	$\geqslant$80% of $TEBC_{sum}$	51,877
	Red Early Warning	$\geqslant$90% of $TEBC_{sum}$	58,361
Off-peak day Reasonable Tourist Capacity	Yellow Early Warning	$\geqslant$70% of $TEBC_{avg}$	24,365
	Orange Early Warning	$\geqslant$80% of $TEBC_{avg}$	27,846
	Red Early Warning	$\geqslant$90% of $TEBC_{avg}$	31,326

As shown in Table 3-8, when the number of tourists reaches 34,000 on peak days in winter half of the year, 45,000 on peak days in summer half of the year and 24,000 on non-tour days, scenic spots will issue warning announcements to the Intelligent Platform. When the number reaches the red warning range, emergency responses should be adopted, such as diverting tourists or batch entry, and measures should be taken to ensure the safety of tourists.

3.3.2 Analysis of the Annual Bearing Capacity

Estimating the total capacity of the Huangshan Scenic Area to receive tourists in the whole year is based on this calculation: 34,807 people/day×365 day = 12,704,555 people/year. According to the *Code for the Planning of Scenic Spots* (GB 50298-1999), the formula for calculating the annual carrying index is as follows:

Annual Bearing Index = Number of visitors received annually/Annual Tourism Environmental Capacity

In 2017, the number of visitors in Huangshan Scenic Area was 3,368,688, with an annual carrying index of about 0.3. According to the current evaluation index, it should be rated as a weak carrying capacity.

3.3.3 Assessment of the Bearing Pressure

(1) The Ecological Load-bearing Pressure Caused by Space Capacity Overload Should Not Be Neglected

Tourist arrival rates at various scenic spots on the mountain are different, resulting in high arrival rates at some scenic spots, which consequently brings environmental pressure and becomes a bottleneck for the capacity of the whole mountain. For example, Beihai now has a basic tourist rate of 100%, and there will be a serious overload of tourists during the peak time. Therefore, the environmental capacity of Beihai will very easily become a bottleneck for the environmental capacity of the whole scenic area. The Yuping and Tianhai scenic areas, as tourist distribution centers due to their cableways, transportation, geographical location and tourist routes, are also overloaded during peak times.

At the same time, during the peak period of tourism, scenic spots with overloaded space capacity, such as Beihai, Yuping, and Tianhai, will also bear a relatively large ecological load. Numerous tourists will become congested and detained, which may lead to the trampling of vegetation along the mountain road, the excessive instantaneous discharge of litter from traveling, a rising risk level for forest fires, and other dangers, which can potentially destroy the natural ecosystem of Mount Huangshan.

(2) The Growth and Decline of Various Kinds of Bearing Capacity

The basic hardware facilities of the Huangshan Scenic Area have been gradually improved, the ropeways at Yuping, Taiping, and Yungu have been opened sequentially, the travelling time of the Xihai sightseeing cable car and the transportation time of the hot spring highway buses have been shortened, and the turnover rate of tourists has accelerated. Furthermore, the direction of passenger flow is also constantly changed, the time, place, and degree of concentration of passenger flow are changed, and the traffic carrying capacity of scenic spots are continually expanded.

At the same time, the rapid increase in the number of tourists has greatly aggravated the environmental pollution. A large amount of solid waste has been produced, which solely depends on manpower to carry it up and down the hills. Therefore, the capacity of sewage disposal at the scenic spots is very limited. The expansion of the accommodation of the water supply facilities can easily lead to a decrease in the vegetation coverage and changes to the soil texture. Garbage, dust, and sewage generated during construction also intensify the ecological pressure at the scenic spots. During the construction and operation of the ropeway, the natural vegetation is affected, as well as tree growth and later plant development, while the laws for groundwater circulation change, and the living habits of animals are disturbed, which consequently reduces the natural ecological environment capacity of scenic spots.

(3) Uneven Carrying Capacity of the Tourism Environment in Each Scenic Spot

The comprehensive utilization level of the environmental carrying capacity for tourism Huangshan Scenic Area is weak throughout the whole year, seriously weak in the off-season, and moderate or slightly overloaded during the peak season, which indicates that the overload problem at Mount Huangshan is caused by an uneven distribution of time and space. Therefore, in order to solve these problems, we should not only expand the carrying capacity for the peak period, but also make use of the remaining capacity of the other scenic spots and alleviate the pressure from tourists in the overloaded scenic spots during the peak period.

Presently, in the eight scenic spots of the Huangshan scenic area, the utilization of the comprehensive carrying capacity at the Beihai, Xihai, Yuping, and Tianhai scenic spots is overloaded, and this overloading will very easily occur during the peak period, which is likely to become the bottleneck for the tourist capacity of the entire Huangshan scenic area. The hot springs and Yungu scenic spots are not easily overloaded by visitors, and thus there is inconsistency between the component carrying capacities. But these two scenic spots have frequent tourist turnover and a low ecological carrying capacity, so ecological restoration projects can be attempted. As for the Xihai Grand Canyon and Songgu scenic spots, there is also an inconsistent diversion of tourists, most of whom do not stay there long enough, and they tend to end their journey by ropeway and sightseeing cable car.

4 The Spatial Division of Biodiversity Protection within the Huangshan Scenic Area

4.1 Target Analysis of Biodiversity Protection within the Huangshan Scenic Area

Mount Huangshan is a picturesque mountainous area with ancient and famous trees as its breathtaking topography. Combined with its current status of being a priority area of the Huangshan-Huaiyu Biodiversity Conservation Area, the objectives of this area's biodiversity conservation should include: ①Rare and endangered wild animals and plants along with their habitats; ②Natural disaster sensitive area status; ③A natural and culturally cognitive landscape with ancient and famous trees as a focus.

4.1.1 Species Habitat

Based on the distribution characteristics of animal diversity resources in the Huangshan scenic area, *Neofelis nebulosa* and *Muntiacus crinifrons* are the key species to be considered for animal diversity protection. Orchidaceae is the representative for rare and endangered plant protection. According to the Niche Theory, the indexes of altitude and ecosystem type are selected to identify the distribution range of the target species. This provides a basis for the accurate protection of wildlife habitat.

4.1.2 Natural Disaster Sensitive Area

Areas with frequent natural disasters are generally areas with fragile ecological environment or greatly affected by human activities in adverse ways. It is of positive significance to identify disaster sensitive areas within the Huangshan Scenic Area for pest control, forest fire prevention, and control. In addition, it also plays an important role in the timely detection of disasters in the scenic area, suppression of diffusion, and adoption of necessary measures.

4.1.3 Natural and Cultural Landscape

The influence of human activities on the living state of biodiversity is determined by cultural values. Culture expresses the interaction between humanity and biodiversity, and culture is also a reliable social force for biodiversity protection. The ancient and famous trees growing in Huangshan are all a precious gift presented to us by nature. They symbolize the historic and cultural core of Huangshan, and they have been the objects of chanting and praising by literati of the ages. Therefore, protecting the ancient and famous trees in Huangshan is also a way of protecting the core of history and culture of Huangshan. Due to the fact that many ancient and famous trees in Huangshan are distributed in tourist scenic spots, it is necessary to not only allow access to them for tourists, thereby promoting Huangshan culture, but also to protect them from being damaged by the very same tourists. Spiritual heritage is particularly important for Huangshan.

4.2 Delimitation of Important Areas for Biodiversity Conservation within the Huangshan Scenic Area

4.2.1 Delineation Method

The delimitation of important areas for biodiversity protection in the Huangshan Scenic Area is helpful in protecting the integrity of the biological ecosystem, habitats of special species, and distribution of important genetic resources. In order to identify important areas, it is necessary to evaluate and identify the biodiversity protection of Mount Huangshan on two levels; these include: ①Evaluating and identifying the importance of the biodiversity protection function, and obtaining the extremely important area of biodiversity protection; ②Evaluating and identifying the habitat of protected species, and obtaining the highly adaptive habitat area of protected species. According to evaluation results, after deducting any overlapping areas, the important areas for biodiversity conservation of the Huangshan Scenic Area are obtained by combining the extremely important areas and high adaptability areas, as shown in Figure 4-1.

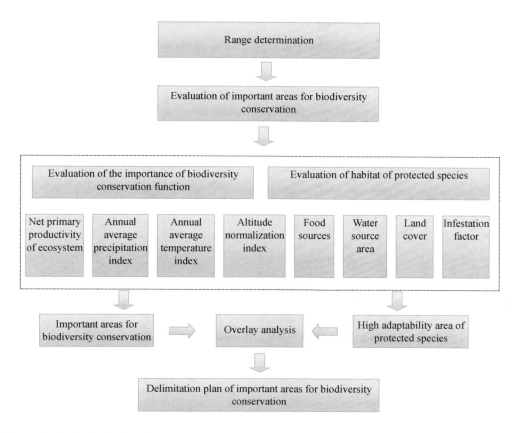

Figure 4-1   Technical road-map for delimitation of important areas for biodiversity conservation in Huangshan

(1) Evaluation of the Importance of Biodiversity Conservation Function

On April 30,2015,the Ministry of Environmental Protection issued the *Technical Guide for Demarcation of Ecological Protection Red Line* (MEP issued 〔2015〕 No.56), which is applicable to the delimitation of ecological protection red line in the People's Republic of China. The delimitation of important areas for biodiversity protection in the Huangshan Scenic Area shall refer to the delimitation method of ecological protection red line, and the evaluation method of biodiversity protection function importance is shown in Table 4-1.

Table 4-1 Evaluation Index of the Importance of Biodiversity Protection Function in Mount Huangshan

Evaluation of the importance of biodiversity conservation function	Method based on habitat diversity $S_{bio}=NPP_{mean} \times F_{pre} \times F_{tem} \times (1-F_{alt})$ S_{bio} biodiversity conservation service capacity index	NPP_{mean} Average value of net primary productivity of ecosystem in the assessment area for many years	NDVI data products synthesized by MODIS (250 m) every 16 days over the years
			Data products provided by high resolution satellites such as Landsat or resource 3 over the years
			Vegetation type distribution map
			Monthly total solar radiation at meteorological station for many years
			Monthly net solar radiation at meteorological station for many years
			Monthly total precipitation at meteorological station for many years
			Monthly total amount of evaporation at meteorological station for many years
			Monthly average ground temperature at meteorological station for many years
			Monthly average temperature at meteorological station for many years
			Monthly average maximum temperature at meteorological station for many years
			Monthly average minimum temperature at meteorological station for many years
			Monthly average pressure at meteorological station for many years
			Monthly average water vapor pressure at meteorological station for many years
			Monthly average wind speed at meteorological station for many years
			Monthly relative humidity at meteorological station for many years
		F_{pre} : Annual average precipitation normalized to 0-1 for many years (more than 30 years)	Annual total precipitation at meteorological station for many years
		F_{tem} : Normalization of multi-year (10-30 years) average temperature data to 0-1	Annual average temperature at Meteorological Station for many years
		F_{alt} :The altitude of the evaluation area is normalized to 0-1	DEM digital elevation model

（2）Species Habitat Identification

The habitat evaluation of protected species is an evaluation method that clearly defines the niche of target species as the principle and ensures the sustainable development of species. Evaluation factors include natural factors and human factors, among which, natural factors include: food source (spatial interpolation), land cover (LUCC), elevation (DEM), slope, slope direction, and water source. Human factors include: residential areas, roads, and mining areas. This method is utilized to not only protect the existing habitats of species, but also to reserve space for the increase of species population within a certain range.

（3）Identification of Landscape Visual Area

The core function of the Huangshan Scenic Area is tourism. It is the most basic requirement to protect the integrity of the viewable landscape of the Huangshan Scenic Area. With this function, we also find the bottom line diversity of the viewable landscape of the Huangshan Scenic Area. Identifying the landscape horizon with the first planned tourist route taken as a node and GIS used to simulate the range of tourists' horizon.

（4）Natural Disaster Sensitive Area

Natural disasters are an important factor in the sharp decline of biodiversity; therefore, prevention of natural disasters is an important means to protect regional biodiversity. By identifying the natural disaster sensitive areas within the Huangshan Scenic Area, the efficiency of risk prevention will be effective and greatly improved. When considering the primary natural disasters faced by Huangshan Mountain, the highly sensitive area to disasters in the scenic area is calculated by means of GIS.

（5）Identification of the Ecological Corridor

From the perspective of biological protection, the ecological corridor is a narrow strip of vegetation for wild animals. This natural asset can generally promote the movement of biological factors between two places. The establishment of an ecological corridor is not only an important method for landscape ecological planning, but it's also an important measure to solve the current landscape fragmentation caused by human activities and many environmental problems. Using the shortest path algorithm and the spatial analysis function of GIS, the animal migration corridor of the Huangshan Scenic Area can be identified and constructed.

4.2.2 Delineation Scheme

(1) Object of Evaluation

On the premise of not affecting the main function of a scenic spot, the protection objectives of different levels can be realized; they are: ①Protecting the integrity of the ecosystem; ②Reserving species habitats; ③No missing of the landscape; ④Management and control requirements for preventing the risk of natural disasters.

Based on the existing remote sensing data and ground monitoring data, a geographic information system is applied to identify ①The biodiversity protection barrier area; ②High adaptability area for species protection; ③Landscape vision area; ④Natural disaster sensitive area; ⑤Ecological corridor where classified management and control measures can be put in place.

(2) Evaluation Process

① Evaluation of the Importance of Biodiversity Conservation Function

Through GIS software, the service value of the biodiversity protection function is classified into four levels by quantile function at the Huangshan Scenic Area level. According to the value of ecosystem services, the four levels are general importance, medium importance, important, and extremely important. The important areas are extracted to form the biodiversity protection barrier area, as shown in table 4-2.

Table 4-2 Evaluation method of the importance of the biodiversity conservation function on Mount Huangshan

Evaluation Item	Evaluation Model	Description of Model Factor Calculation
Importance of biodiversity conservation function	$S_{bio}=NPP_{mean} \times F_{pre} \times F_{tem} \times (1-F_{alt})$ S_{bio} is the service capacity index of biodiversity conservation	NPP_{mean} is the average value of net primary productivity of ecosystem in the assessment area for many years F_{pre} is the interpolation and normalization of annual average precipitation data for many years (more than 30 years) to 0-1 F_{tem} is the interpolation and normalization of mean temperature data for many years (10-30 years) to 0-1 F_{alt} is the altitude normalization of the evaluation area to 0-1

According to the evaluation results, the important areas of biodiversity protection within the Huangshan Scenic Area are typically distributed throughout the vast majority of the scenic area, which covers an area of 144 square kilometers and accounts for 89.67% of the total scenic area. When we compare the importance of different biodiversity conservation areas in Huangshan City, the Huangshan Scenic Spot is an important functional area of biodiversity conservation in the northwest of Huangshan City.

② Adaptability Analysis of Protected Species

According to the Niche Model, GIS technology and the analytic hierarchy process were used to select *Neofelis nebulosa* and *Muntiacus crinifrons* (Column 4-1, Column 4-2) as the research objects of the important biodiversity protection area of Huangshan Scenic Area. These species were used to determine the high degree of adaptability to the area.

Column 4-1　*Muntiacus crinifrons*

Muntiacus crinifrons, also known as Wujin muntjac and Pengtou muntjac, is a species of large muntjac. It is a national level protected wild animal listed in Appendix I of CITES and rated as "vulnerable" by IUCN. It has a body length from 100-110 cm, a shoulder height of about 60 cm, and a weight between 21-26 kg. The hair color of the upper body in the winter is dark brown, and the brown component of the hair color is increased in the summer. Its tail is longer, generally more than 20 cm. Its back, belly, and dorsal side of its tail is black. The underside and edges of the anterior side of its tail are pure white, which is very eye-catching.[①]

---

① http://www.jiaodong.net/news/system/2009/06/16/010555671.shtml.

The *Muntiacus crinifrons* is a special animal in China; it does not differentiate into subspecies, and its distribution range is very narrow. It is distributed in the south of Anhui, west of Zhejiang, Huaiyuan in the east of Jiangxi, and Mount Wuyishan in the north of Fujian. It inhabits evergreen, broad-leaved forests, mixed evergreen and deciduous broad-leaved forests and near shrubs on the mountain with an altitude of about 1,000 m.

Muntiacus crinifrons are timid and have a strong sense of fear. They usually move in the morning and dusk. During the day, they often rest near the roots of trees or in caves. If there is a little noise, they immediately run into the bushes and hide. They will follow a relatively fixed route in steep places and often trample out a 16-20 cm wide path, but they do not have a fixed route in flat, uniform topographical places.

Column 4-2 *Neofelis nebulosa*

Neofelis nebulosa is a mammalian feline with four subspecies. As the smallest one in pantheriinae, it has a body length of about 70-110 cm, a tail length of about 70-90 cm, and a weight of about 16-40 kg. *Neofelis nebulosa* is distributed in Southeast Asia. Its habitat starts from the westernmost part of Nepal and extends eastward to Taiwan of China including Myanmar and the south of Mount Qinling in China. In the south, its habitat starts from the east of India and the Indochina Peninsula. From there, it extends southward to the Malay Peninsula and beyond. Clouded leopards are a highly arboreal species. They can often be found resting and hunting on trees, but they actually spend more time on the ground than in trees. Clouded leopards live in subtropical and tropical hill and mountain evergreen forests, most often in evergreen tropical virgin forests, but they can also be seen in other habitats, such as secondary forests, mangrove swamps, grasslands, shrubs, and coastal broad-leaved forests. Typically these areas have a vertical height of 1,600-3,000 m above sea level and an environmental temperature of about 18-50 ℃.①

---

① http://tupian.baike.com/ipad/a1_55_75_01300000174719121437758683327_jpg.html.

According to the target species living habits analysis, there are eight ecological factors affecting the habitat distribution. These are divided into natural factors and human factors. Natural factors include: food sources, land cover, altitude, slope, slope direction, and water sources. The human factors include: residential areas and roads. If we investigate previous, relevant literature protection expert inquiries, we can notice key factors, like the best altitude for the protected species, the best distance from water sources, the variations in different land cover types, the habitat slope, the slope direction, and the distance from human interference (settlements, roads, etc.).

When we observe the contrasting ways of utilizing the above eight ecological factors (variables) by the protected species, different values (value range: 0-100) are given respectively. An area with a score of 100 in each ecological factor would be the best habitat.

With the help of the spatial analysis function of GIS, each factor is superposed according to weight value to get the best habitat distribution map of the protected species. The calculation results are classified to extract the high adaptability area of the protected species. According to the analysis results, the highly adaptive area of *Muntiacus crinifrons* is scattered, and the habitat is obviously affected by roads. The "extremely adaptive" area accounts for 33.62% of the total scenic area. Typical distribution is in the east and west of the scenic area. These areas have a large and complete habitat, among which include the central part of the Fuxi management area, the southern part of the Diaoqiao management area, the western part of Songgu management area, and the western part of Fugu management area. These areas are the key nodes to carry out *Muntiacus crinifrons* monitoring and local protection in the future. The forest ecosystem in this area is complete, the water source is sufficient, the distance from tourists is far, and human interference is small. These characteristics are conducive to the restoration of the *Muntiacus crinifrons* population.

A review of the analysis results shows that the extremely adaptive area of *Neofelis nebulosa* is concentrated in the central and western regions, and the area of the extremely adaptive area accounts for about 8.92% of the total area. The forest ecosystem in this area is complete, the water source is sufficient, the distance from tourists is considerable, and human disturbances are small. These are conducive conditions to the restoration of the *Neofelis nebulosa* population. Another look at the analysis indicates that the habitat of *Neofelis nebulosa* is obviously affected by roads. The central part of the Fuxi management area, eastern part of Diaoqiao management area, Yuping management area, and other areas have been found to be extremely adaptive areas for *Neofelis nebulosa*. They exhibit a high probability of possible activities in these areas, which are important nodes for local protection and the monitoring of *Neofelis nebulosa*.

Again looking at the analysis, the Huangshan Scenic Area benefits from a high forest coverage, and the original vegetation is completely preserved. These conditions have achieved good results in the protection of rare and endangered plants, like those represented by Orchidaceae. The "extremely adaptive area" of rare and endangered plants is large; it accounts for 31.91% of the total scenic area. Orchidaceae are primarily distributed in the southern part of the hot spring management area, the northern part of the Yanghu management area, the eastern part of the Fugu management area, and the southwestern part of the Fuxi management area. Additionally, the protection measures utilized and embodied from the "one tree, one policy" have been implemented within the Huangshan Scenic Spot. Effectively, this protects the famous ancient trees located near tourist roads and completes the ancient tree resources (Figure 4-2).

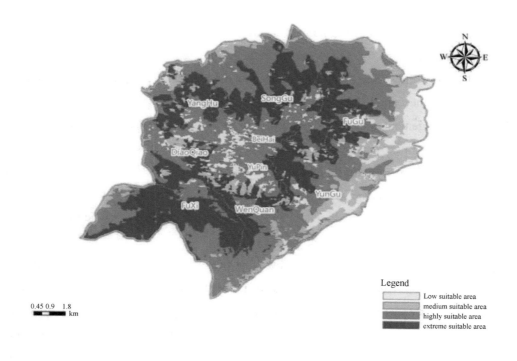

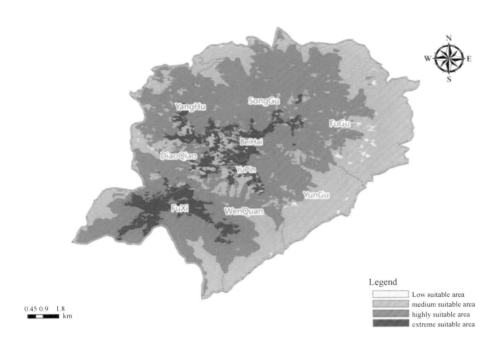

Figure 4-2　Key Species Habitats and Adaptability Evaluations in Huangshan Scenic Area

（3）Landscape View Analysis

Based on the DEM, tour route map, and other scenic area planning data of the Huangshan Scenic Area, the scenic area that tourists are able to view while travelling on foot is simulated. The visual area of the scenic area is then identified to form the landscape viewing area. The total area is about 92.86 square kilometers, and it accounts for 57.82% of the total scenic area. The identification area mainly encompasses the Beihai management area, Songgu management area, Yuping management area, Yungu management area, and the hot spring management area. In addition, the Huangshan Scenic area covers the main scenic spots, field exploration area, and other scenic spots.

（4）Sensitivity Analysis of Natural Disasters

The Huangshan Scenic Area is a strong cutting area with a cutting depth of 500-1,000 m, precipice, convex hillside, steep slope, and dangerous peak. The steep slope varies between 60°and 74°, and the soil layer is thin. These conditions make for an area which is sensitive to water and soil loss. Moreover, the analysis results show that this fragile soil and erosion area（Figure 4-3）, is an extremely sensitive area that also accounts for about 8.67% of the total area. It is found in the Yuping management area, Beihai management area, and Yungu management area.

Figure 4-3 Sensitivity Analysis of Natural Disasters in Huangshan Scenic Area

The main vegetation type found at Huangshan is coniferous forest, which is also an important landscape subject of Huangshan. It is vulnerable to Dendrolimus and

Bursaphelenchus xylophilus. According to the principle of prevention first and the existing data, and through spatial calculations, the Dendrolimus disaster sensitive area has been identified The extremely sensitive area accounts for about 14.52% of the total scenic area, and the sensitive Dendrolimus disaster area located in the Huangshan Scenic Area is mainly located in the low-altitude Yungu and Fugu management areas.

The staple vegetation type of Huangshan is the coniferous forest, and forest fires are the key disaster we must prevent. Data from the Huangshan Scenic Spot has been able to identify the first, second, and third class fire sensitive areas. The first level area can be essentially found around the core scenic area.

4.2.3 Integration of Important Regional Schemes

By examining the evaluation results, deducting the overlapping areas, linking up the existing *Master Plan for Tourism Development of the Huangshan Scenic Area* along with other plans, combining the extremely important areas and high adaptability areas, the important biodiversity protection areas of the Huangshan Scenic Area can be obtained (Figure 4-4), as shown in Table 4-3.

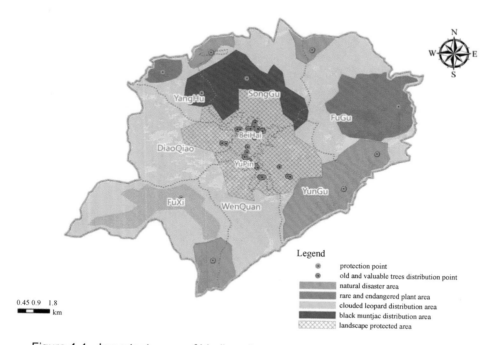

Figure 4-4 Important areas of biodiversity protection in Huangshan Scenic Area

Table 4-3 List of important biodiversity conservation areas within the Huangshan Scenic Area

	Category of Important Area		Distribution Range	Area/km²
I	Biodiversity protection barrier area		Most locations within the scenic area	134.78
II	High adaptability area of protected species	Adaptive area of *Muntiacus crinifrons*	Songgu, Fugu, Yungu, and Fuxi management areas	60.80
		Adaptive area of *Neofelis nebulosa*	Fuxi and Diaoqiao management areas	35.20
		Adaptation area of endangered plants	Fugu and Yungu management areas	56.00
III	Scenic landscape area		Songgu, Beihai, Yuping, Yungu, and hot spring management areas	85.43
IV	Vulnerable natural disaster area		Beihai, Yuping, Fugu, and Yungu management areas	91.78

4.3 Management and Control Measures for Biodiversity Conservation within the Huangshan Scenic Area

4.3.1 Biodiversity Protection Barrier Area

It is distributed in Most natural ecosystem areas of a scenic spot contain a biodiversity protection barrier area of a substantial size. Within this area, the integrity of the ecosystem must be ensured, and the ecological service must not decline. Regular assessment should be carried out to reduce the interference from human activities.

4.3.2 High Adaptability Area of Protected Species

The high adaptability area of protected species relates to a precious habitat for the protection and restoration of rare and endangered species. The high adaptability area of *Muntiacus crinifrons* can primarily be observed in the Songgu, Fugu, Yungu, and Fuxi management areas. Additionally, the high adaptability area of *Neofelis nebulosa* can be found in the Fuxi and Diaoqiao management areas. Moreover, the high adaptability area of endangered plants is typically existent in the Fugu and Yungu management areas. By definition, the high adaptability area must be far away from human activity areas, tourists must be strictly prohibited from entering, and great care must be taken to prevent human construction activities from disturbing animals in these habitats. Species investigation and target species monitoring should be carried out in high adaptability areas. The population status of target species(*Muntiacus crinifrons*, *Neofelis nebulosa*, etc.)should be evaluated regularly. High adaptation areas must be opened and closed with respect to natural laws, the different seasons, and day and night changes to protect biological resources.

4.3.3 Landscape View Area

The landscape view area is the principle landscape area for tourists visiting Huangshan; this area has a profound effect on tourists' perceptions. This area is predominately evident in the Songgu, Beihai, Yuping, Yungu, and hot spring management areas. In this area, we must pay attention to the integrity of the natural landscape and avoid building an incongruous man-made landscape. In the last planning stages of this area, we need to focus on a harmonious relationship between construction and the landscape.

4.3.4 Natural Disaster Sensitive Area

The natural disaster sensitive areas identified in the Huangshan Scenic Area are mainly faced with the risk of soil erosion, lightning, forest fires, and forest insect pests. This area is usually conspicuous in Beihai, Yuping, Fugu, and Yungu management areas. When considering the forest fire disaster caused by lightning in the Piedmont area, an effective fire early warning and protection system should be implemented. In addition, prevalent diseases and insect pests in the scenic spot should be monitored regularly, and protection from these dangers should be given priority.

4.3.5 Time Control

According to the management requirements of *Regulations on Closed Rotate Rest of Scenic Spots in the Huangshan Scenic Area-Basic Requirements*（DB 34/T 1241-2010）, the Rotate Rest System will continue to be implemented in the developed areas of the Huangshan Scenic Area. By way of artificial measures to assist in natural recovery, we can restore undergrowth vegetation, promote tree growth, improve the ecological environment, and realize the sustainable development of the entire scenic area. At the same time, it is necessary to focus on the seasonal alternative migration of animals, adjust time control measures according to the periodic migration activities, and achieve coordinated development and protection.

5 Problems of Biodiversity Protection in the Huangshan Scenic Area

5.1 Uneven Distribution of the Environmental Bearing Capacity of Scenic Spots, a Natural Ecology under Immediate Pressure

Based on the static model for tourism environmental capacity and the development and construction planning for the scenic spot within the given time period (2018-2030), the tourism environmental capacity of the scenic spot is calculated and analyzed from the spatial, ecological, and economic environments, and from the socio-psychological environmental capacity, and the main conclusions are as follows:

5.1.1 The Core Scenic Spot is Prone to Overload during the Peak Period

At present, the annual comprehensive utilization level of the tourism environmental capacity of the Huangshan scenic area is under a weak load, but the tourism environmental capacity of each scenic spot is different. To be specific, the tourism environmental capacity of Beihai, Xihai, Yuping, and Tianhai scenic spots are generally lower than that of Donghai, Diaoqiao, and other scenic spots, and the utilization of the comprehensive bearing capacity is easily overloaded. The basic tourist rate at the Yuping scenic spot is 100%, and during the peak period, there will be a large number of tourists visiting, and thus it easily becomes the weak link for environmental capacity in the whole scenic area. Due to the dense cable way, traffic and lines, the core scenic spot becomes a tourist distribution center, and it is easy for tourists to jam up and linger during the peak period.

5.1.2 The Immediate Pressure of Tourist Scale on the Natural Ecological Protection in the Scenic Area

During the peak period, a sudden surge occurs in the number of tourists at the core

Huangshan scenic spots, and the lack of effective guidance and crowd dispersal easily leads to lines of traffic and congestion along tourist footpaths at the scenic spots, which may also cause vegetation trampling and damage along the mountain road. In addition, there is a rapid increase in the discharge of tourism waste, a rising risk of forest fires, and other related dangers, which form a potential threat to the natural ecosystem of Huangshan. In addition, the infrastructure and cableways reach their upper limit for transportation, there is a huge discharge of solid waste and sewage, and the overloading of the water and power supply further aggravate the degree of disturbance to the natural habitat, thus greatly affecting the ecological protection value of the natural ecosystem and the landscape integrity and biodiversity of Huangshan.

The overload problem faced by the Huangshan scenic area is mainly reflected in the uneven distribution of environmental capacity. Therefore, it is necessary to focus on how to disperse the pressure from the tourists in the overloaded scenic spots during the peak period of tourism, thereby protecting the natural ecological capital from too much interference, while still providing high-quality eco-tourism services.

5.2 The Tough Task of Biodiversity Management and Protection, Weak Monitoring and Early-Warning Capacity

5.2.1 Unclear Biodiversity Background Data

The background data for biodiversity in the scenic area has not yet been systematized. The lack of basic data for the clouded leopard, golden cat, civet, jackal, badger, and other national and provincial key protected animals increases the difficulty of the conservation work. There is an urgent need to collect the number, distribution area, habitat, and other information for the species at the top of the food chain in the scenic area. Moreover, it is difficult to observe rare birds such as the barred-backed pheasant, the background of amphibious reptiles such as the giant salamander, rana spinosa, big-headed turtle, and ancistrodon acutus is unclear, and there is no basic data for the insect and microbial resources. In addition, the general key species, special crops and other characteristic biological resources in the scenic area need to be systematically investigated and studied.

5.2.2 The Severe Situation Caused by the Pinewood Nematode and Alien Species Invasion

The survival of *Pinus taiwanensis* is threatened by the pine wood nematode. The Huangshan scenic area is vulnerable to the occurrence of this disease in every aspect. Once the pine wood nematode disease breaks out in the Huangshan Scenic Area, it may cause devastating damage to the human and natural landscape of the area, and at any time, it may cause great and irreparable harm to both the area's ecological resources and to the precious natural human heritage in the Huangshan scenic area.

It has also been found that there are 31 species of invasive plants in the scenic area, which are widely distributed and cause certain amount of harm. The populations of garden goldenrod, fall webworm, and other exotic plants have shown an obvious high-growth trend.

5.2.3 Weak Monitoring and Early Warning Capacity for Disasters

The ecological environment of the scenic area is fragile, natural disasters are frequent and destructive, forest fires easily occur, and there is no effective integration of the background data on the resources. It is necessary to further the systematic integration of the data on rare and endangered species in the scenic area, including population number, distribution location, growth status, and other metrics, while also investigating the changing trends in the ecological environment of the scenic area, under the influence of various social and economic factors.

5.2.4 The Relative Lack of Professional Management, Protection Personnel, and Technical Means

The most limiting factor for biodiversity conservation and management in the scenic area is the lack of appropriate technical talent. Specifically, the operation personnel for professional equipment and instruments have insufficient ability, the equipment is facing difficulties in both long-term operation and maintenance, and no effective joint force has been formed. The marketing mode of the publicity for ecological environmental protection and the education for domestic and foreign tourists are relatively simple and shallow, and there is a lack of advanced and developed environmental publicity and education venues and facilities, and the work experience in this aspect remains insufficient. Thus, there is a need for targeted professional skills training, active participation in international exchanges, and learning from more advanced international experience.

5.3 The Unrealized Sustainable Utilization of Resources, the Low Efficiency of Agglomerated Community Development

The total economic value of the biodiversity in the Huangshan scenic area is very high, which is divided into three parts: the direct development value, ecological protection value, and future existence value. However, the main problems and challenges listed below still occur.

5.3.1 Tickets, Catering, and Other Direct Income Account for a High Proportion of the Total Income

Tickets and accommodation income account for an absolute proportion of the direct development value of the biodiversity in the Huangshan scenic area. The profound historical and cultural value of biodiversity in Huangshan has not fully materialized, and the added value to the product is low. If the direct development value of biodiversity in the scenic area only depends on ticket revenues for a long time, the only way to increase revenue is to rely on increasing the number of tourists. This increasing number of tourists will consequently put huge pressure on both the ecological environment and the tourist experience at the scenic spots, which is not conducive to their sustainable development.

5.3.2 Lack of Driving Capacity for Scenic Area Development in the Surrounding Communities

At present, the scenic spot is mainly for sightseeing, a relatively singular function, and it is difficult to cover and drive the development of the related ancillary industries. The lack of opportunities for tourists to participate and experience in specialty shopping, local culture, cultural tours, and other aspects has resulted in fewer repeat visitors and a low revisit rate. In addition, the scenic spot lacks both overall planning and systematic planning for the development and utilization of its biological resources. Most of the residents in the communities surrounding the scenic area are engaged in self-made/self-sold and other sporadic retail activities, which have low profit margins, a high market risk impact, and low development efficiency. Furthermore, the decentralized management can easily cause disorderly development and the waste of biological resources, which is not conducive to protecting either the ecological environment or animal and plant resources.

5.4 Lack of a Traditional Regional Culture and Brand Value Effect

Mount Huangshan has a huge reserve of biological resources and the unique added value of its geography and culture. However, the characteristic local resources, agricultural products, and related cultural products all vary in their type and quality, with a lack of uniform and formal packaging and trademarks, and as such they have not yet been established on an industrial scale. It is thus urgent to enhance the brand agglomeration effect for Huangshan (Huizhou) industry. Thus, one of the problems that we need to pay more attention to is taking advantage of the characteristic regional culture of Huizhou, since the traditional culture is weakening, especially the ability to inherit and the inheriting of biological resources.

We should fully utilize the great natural and cultural heritage value of Huangshan's biodiversity to arouse a strong sense of cultural identity in the public. We need to innovate the means of transmitting the heritage of Huangshan (Huizhou) culture, and establish a sense of mission among the public to protect the natural ecology and biological resources of Huangshan, with the help of cultural carriers.

6 Action Plan for Biodiversity Conservation within the Huangshan Scenic Area

6.1 Protection Objectives and Main Tasks

6.1.1 Overall Objectives and Phased Objectives

In accordance with the requirements of building a "beautiful Huangshan" with "the best ecology, our overall objectives include providing for the most beautiful environment possible, the best protection for plant, animal, and land resources. We must also include best management practices, the highest quality of design and efforts for the most beautiful brand." Our objectives must be based on the strict conservation of biodiversity, with the improvement of ecosystem services as their core, with the promotion of sustainable development of scenic spots as the network, and with the priority action of biodiversity protection as the starting point. These guidelines and standards will ensure that the ecosystem services of scenic spots do not decline, the area of protected space is not reduced, the coordination and unity of biodiversity protection and the economic and social development of the scenic spot is promoted. It will be then that we may finally realize the harmonious coexistence of humanity with nature.

(1) Recent goals

By 2022, the capacity of biodiversity protection will be greatly improved. The background investigation and assessment of biodiversity in panoramic area systems will be completed. The biodiversity monitoring and early warning network system will be fundamental established, and daily monitoring of biodiversity will be carried out at all stations. Local protection and *ex-situ* protection will be strengthened, and a number of conservation communities and biological species resource databases will be built. The management and protection capacity of reservoirs, nurseries, and scenic spots will be greatly upgraded. Biodiversity protection related laws and regulations will be improved to make the Huangshan Scenic Area a national park pilot unit.

（2）Medium Term Goals

By 2025,biodiversity conservation capacity will be significantly enhanced. Numerous kinds of ecosystems, all national and provincial key conservation, and rare and endangered wild animals and plants will be effectively protected. The sustainable utilization capacity of biodiversity will be continuously strengthened to achieve the benefit of sharing between biodiversity protection stakeholders. Domestic and international cooperation will be widely developed, and international popularity and influence will be greatly improved. The publicity, education, and science and technology of biodiversity conservation will be popularized, and the public will become more actively involved in biodiversity conservation.

（3）Long Term Goals

By 2030,biodiversity will be effectively protected. Various management systems and protection policies will achieve remarkable results. Biodiversity management and protection ability and scientific research level will be greatly improved. The ecosystem will be in a good condition. The species resources in the panoramic area will be effectively protected and the genetic resources will be effectively preserved. These benefits will then meet the needs of the sustainable development of the scenic area. The public will consciously participate in biodiversity protection and realize the harmonious coexistence between humanity and nature.

6.1.2　Main Tasks of Biodiversity Conservation

（1）Carrying out Spatial Division and Constructing the Distribution Pattern of Important Regions

Establishing the delimitation plan for the important areas of biodiversity protection in Mount Huangshan. In accordance with the *Technical Guide for Demarcation of Ecological Protection Red Line*（MEP issued〔2015〕No.56）issued by the Ministry of Environmental, Protection the important functions of biodiversity protection, protected species habitat, landscape and natural disaster sensitive areas of the scenic spot have been identified and evaluated. This step has been done so as to protect the ecosystem completely, reserve the habitat of important species, keep the landscape intact, and effectively prevent the risk of natural disasters.

Building the spatial distribution pattern of important regions. Based on the *Master Plan of the Huangshan Scenic Area*（2007-2025）, the important areas for biodiversity protection in the Huangshan Scenic Area have been demarcated, the scenic area spatial pattern of biodiversity protection will be constructed, and the integrity of the ecological system, tourism ecological environment safety, habitat of special species, and distribution of important genetic resources shall be wholly protected.

(2) Determining the Protection Objectives and Implementing Classified Management and Control Measures at Different Levels

Setting protection objectives and implementing hierarchical control. Based on the guiding ideology of "controlling according to the red line, and leaving space in advance," the spatial pattern of biodiversity protection in the scenic area is managed and controlled at different levels. The protection key nodes, such as important species, their habitats, fragile ecology, and sensitivity to natural disasters are taken as the first level, red line area; the biodiversity protection barrier area as the second level, core area; and the landscape horizon belt as the third level, optimized development area.

Determining the bearing red line and carrying out classified management and protection. Control red lines shall be set for key conservation nodes such as important species and their habitats, fragile ecology, and sensitivity to natural disasters. The purpose of this is to strictly limit resource development and human activities. This will help to ensure against the protected area decreasing. Damaged ecosystems in the biodiversity protection barrier area shall be repaired naturally or artificially to ensure that the ecological function does not decrease. Four landscape horizon optimization zones shall be optimized for development. These zones will reserve space for future development. Red lines shall be set for tourists' bearing and environmental consumption to ensure that the use of resources does not exceed the line.

(3) Scientific Utilization of Resources and Joint Development of Surrounding Communities

Promoting the protection and utilization of biological resources. Promoting the protection and sustainable utilization of biological and human resources related to biological diversity in the Huangshan Scenic Area based on the principle of "making the best of everything and focusing on the future." For example, the electronic identification of rare and endangered endemic species, such as *Pinus taiwanensis*, *Rhododendron*, and *macaque*, may be utilized to improve the construction of an *ex-situ* breeding base and genetic resource bank. The standardized management and protection of the original species of important and promising biological resources in Huangshan, such as tea, fungus, bamboo shoots, medicinal materials, and other scenic spots can be organically combined with the regional characteristic cultural resources to establish a scientific market development mechanism. Measures such as banning the exploitation of fungi can reduce the degree of over exploitation of this resource. The prevention and control of alien invasive species and technical research should also be actively carried out.

Participating in collaborative development with surrounding communities. The scenic area should be coordinated with the surrounding "five farms and one town" so as to jointly build the characteristic tourism brand of the Huangshan Scenic Area, and jointly promote the construction of Huangshan "grand tourism" industry. By highlighting the geographical indications and regional characteristics of Huizhou culture and "world natural and cultural heritage, " we will establish the Huangshan Tourism Ecological trademark and service mark, extend the tourism industry chain, integrate the traditional cultural knowledge of Huizhou architecture, folk arts, crafts, featured products, Huizhou diet, etc. We will then be able to realize a "unified trademark, uniform, unified service, and coordinated management" from the top to the bottom of the mountain. We will be able to highlight the advantages and benefits of the Huangshan Tourism Brand to realize the coordinated, win-win development of the scenic area and surrounding communities.

(4) Promoting Tourism Publicity and Creating a Unique Ecological Brand of the Scenic Area

Strengthening public awareness and educating them on the biodiversity protection in scenic spots. The scenic area should promote publicity through multiple types of media. We must improve public awareness of ecological protection. We must establish mechanisms for information disclosure, public supervision, and reporting. We must improve ways for the public to participate in ecological protections. We must request and support travel agencies to strengthen the publicity and education of tourists, establish and improve the multilingual guide system of the scenic spot, and display the natural and cultural history, animal and plant resources, cultural and artistic works of the scenic spot to the international and domestic public. We must vigorously promote Huangshan cultural products, creative products, and ecological goods through scenic spot stores, and online at WeChat, Weibo, Internet e-commerce, and other channels.

Deepening the partnership for biodiversity conservation. The scenic area should develop and deepen domestic coordination, international cooperation, information exchange, attract corresponding international and domestic relevant departments, scientific research institutions, non-governmental organizations, enterprises and institutions, communities, the public, and other forces to participate in the organization and implementation of biodiversity protection projects. We must attach great importance to the construction of teaching and scientific research, natural education and other bases with colleges and universities, primary and secondary schools, cultivate long-term developmental partnerships, and establish the Huangshan Scenic Area ecological tourism brand.

(5) Making Special Plans to Broaden the Way of Resource Protection and Utilization

Developing a priority action plan for biodiversity conservation in Huangshan. Based on the *China*'s *Biodiversity Conservation Strategy and Action Plan* (2011-2030) and the *Master Plan of the Huangshan Scenic Area* (2007-2025), the needs of biodiversity protection in Mount Huangshan, and different protection objectives and priorities, priority action plans for biodiversity protection at Mount Huangshan have been formulated. All protection plans have been defined and implemented in batches. They have also been given primary and secondary designations.

Taking various ways to promote biodiversity protection. These include strengthening the restoration of a degraded ecosystem, population expansion of rare and endangered species, *in-situ* and *ex-situ* conservation, and scientific development and utilization of species resources, building a dynamic monitoring and early warning platform to promote species resource information management, emphasizing the coordinated development of scenic spots and surrounding communities, paying attention to the construction of professional and technical personnel training mechanism, establishing all-round and multi angle biodiversity protection and management strategies, and ensuring the sustainable development of the Huangshan regional ecological security and tourism industry.

6.2　Priority Actions for Biodiversity Conservation within the Huangshan Scenic Area

According to the strategy and stage objectives, 10 priority areas and 30 priority actions for biodiversity conservation in the scenic area have been determined.

Priority Area I: Biodiversity Survey, Monitoring and Information Management

Action 1: Biodiversity Background Survey, Cataloguing and Assessment

①Systematically carrying out the background comprehensive investigation of key species resources in the scenic spot, and establishing resource archives and catalogues.

②Investigating the vegetation status and drawing up a vegetation distribution map based on the "3 S" technology and field investigations.

③Assessing the distribution pattern, change trend, protection status, and existing problems of important ecosystems and biological groups in the scenic spot. A comprehensive assessment report must be issued.

④Carrying out the evaluation of ecosystem services and economic value of species resources in scenic spots, and exploring the establishment of an evaluation system.

Action 2: Investigation and Cataloguing of Resources of Characteristic Biological Species

①Carrying out investigations, collection and arrangement of livestock, poultry and aquatic product resources in the scenic spot and surrounding communities, and establishing resource files and catalogues.

②Carrying out special investigations on the germplasm resources of important trees, wild flowers, medicinal organisms, wild vegetables, agriculture and related plants in the scenic area, species determination, quantity, distribution and habitat status of the species resources, collecting and preserving the resources, catalogue and building the database.

Action 3: Survey and Cataloguing of Traditional Knowledge Related to Biodiversity Conservation

①Carrying out the investigation, recording, sorting out, and cataloguing of traditional knowledge related to biodiversity protection in the communities around the scenic spot. Its customs and beliefs can be used as important management resources for the biodiversity protection and sustainable development.

②Focusing on investigating and sorting out the traditional geographical products, traditional technologies, and usages related to biodiversity used by local communities.

Action 4: Biodiversity Monitoring and Early Warning

①Carry out the monitoring and early warning of national and provincial key protected species, mainly including the national level Ⅰ and level Ⅱ protected animals and plants distributed in the scenic spot and the protected animals and plants in Anhui Province. Carrying out monitoring and early warning of risks to habitat or population and its habitat together.

②Carrying out the monitoring and early warning of the unique rare and endangered plants and their habitats in the key areas of Mount Huangshan. Monitoring and early warning of habitat and community dynamics, studying the internal mechanism and external factors of threatened species, and finding new threatened species in time to intercede.

③Carrying out monitoring and early warning of important ecosystems. Selecting a representative forest ecosystem to carry out monitoring and early warning in key areas of protection, grasping changes in time, and carrying out protection and management scientifically.

④ Carrying out monitoring and research on the impact of tourism project construction on biodiversity. Carrying out long-term monitoring of tourism activities in the scenic area to analyze the interference of tourism development on important ecosystems and key protected species of animals and plants in the scenic spot.

⑤Improving infrastructure, purchasing advanced equipment, and gradually establishing a digital biodiversity monitoring and early warning network system to share data.

Action 5: Information Management of Biological Species Resources

Establish the biodiversity information management system of Huangshan Scenic Area. Collecting and integrating the existing biological resource information, supplementing the key investigation, and establishing the species resource database including species classification, distribution area, resource reserves, protection level, and threatened and protected conditions.

Priority Area II: Protection of Rare and Endangered Wild Species Resources

Action 6: Protection of Macaques and their Habitats in Huangshan

①Carrying out a special investigation on the number, distribution area, and habitat environment of short tailed monkeys in Huangshan, establishing the ecological niche of short tailed monkeys in Huangshan, and analyzing the factors restricting their growth and reproduction according to their living habits.

②Setting up a protection area for short tailed monkeys in Huangshan to ensure the population number of wild short tailed monkeys.

③Carrying out the research on the artificial breeding and reproduction of short tailed monkeys in Huangshan, so as to obtain a good result in the early stage of release into the wild.

Action 7: Protection of Predators Such as *Neofelis nebulosa*

Focusing on investigating the number, distribution areas, and habitat conditions of *Neofelis nebulosa*, *Catopuma temminckii*, *Viverra zibetha*, *Ursus thibetanus*, *Cuon alpinus*, *Meles meles* and other key protected animals at the national and provincial levels in Anhui Province, and analyzing the factors restricting their growth and reproduction according to their living habits; setting up protection communities and biological corridors, establish conventional protection systems, and formulate protection programs.

Action 8: Protection of *Syrmaticus Ellioti*, *Lophura Nycthemera*, and Other Birds

Focusing on investigating the number, distribution areas, and habitat conditions of key protected birds on the national and provincial levels, such as *Syrmaticus ellioti*, *Lophura nycthemera*, and *Pucrasia macrolopha*. Setting up conservation communities and biological corridors, establish conventional conservation systems, and formulate conservation plans.

Action 9: Protection of *Amphibians* and Reptiles Such as *Andrias* and *Deinagkistrodon Acutus*

Focus on the population investigation, distribution area, and habitat condition of the key protected animals in China and Anhui Province, such as *Andrias*, *Quasipaa spinosa*, *Platysternon megacephalum*, and *Deinagkistrodon Acutus*. Analyze the factors restricting their growth and reproduction according to their living habits. Set up protection community, establishing conventional protection system, and formulating protection scheme.

Action 10: Protection of *Muntiacus Crinifrons* and Other Ungulates

Focusing on investigating the number, distribution area, and habitat of key protected animals at national and provincial levels in Anhui Province. These would include *Muntiacus crinifrons*, *Hydropot*, and *Capricornis sumatraensis*, Analyze all factors restricting their growth and reproduction according to their respective living habits. Establish conservation communities and corridors, establish conservation systems and formulate conservation plans.

Action 11: Protection and Population Expansion of Rare and Endangered Plants Such as *Kirengeshoma Palmata*

Focusing on the protection of rare and endangered plants, such as *Kirengeshoma palmata*, *Emmenopterys henryi*, *Pseudotsuga gaussenii*, *Pseudolarix amabilis*, *Cercidiphyllum japonicum*, *Magnolia cylindrica*, *Oyama sieboldii*, *Sorbus amabilis*, etc., carrying out the research on conservation biology, protecting the existing plants and populations, carrying out artificial propagation experiments, and gradually expanding the artificial population.

Action 12: Protection and Population Expansion of Rare Orchidaceae such as *Cypripedium Japonicum*

Carry out special investigations on wild resources of Orchidaceae on Mount Huangshan, represented by *Cypripedium japonicum*, *Cephalanthera falcata*, and *Cephalanthera erecta*. Set up special field protection points, carry out artificial breeding research, carry out field playback test, and protect germplasm resources.

Priority Area III: Strengthening Ecosystem Restoration

Action 13: Restoration of Degraded Ecosystems

Restore the forest ecosystem, river, and other freshwater ecosystems which are seriously disturbed by human activities, diminished in their service capacity, and difficult to restore naturally. Formulate a recovery plan, and carry out biodiversity recovery construction according to local conditions in accordance with the principle of minimum risk and maximum benefit. Combine engineering measures with biological measures.

Action 14: Restoration of Pure Plantation Ecosystems

Restore the biodiversity of the pure plantation ecosystem such as pine, fir, and bamboo with single species, simple community structure and low ecological function. Introduce and cultivate local species, and promote the natural recovery through artificial measures, so as to improve the biodiversity and ecological function of the pure plantation ecosystem.

Priority Area IV: Strengthen the Control of Invasive Alien Species

Action 15: Control of Invasive Alien Species

①Increase investment in measuring and reporting points and quarantine checkpoints for invasive alien species. Strengthen routine quarantine supervision over invasive alien species, and strictly enforce quarantine declaration and entry inspection to prevent human transmission of disease.

②Work out the prevention and control plan of alien invasive species. Strengthen the prevention and control technology research of *Bursaphelenchus xylophilus*, *Dendrolimus*, *Monochamus alternatus*, and other alien invasive organisms. Focus on harmless treatment of dead pine trees.

Priority Area V: Strengthening Local Protection and *Ex-Situ* Protection

Action 16: Improving the Management Level of Local Protection at Scenic Spots

Improve the organization, set up a special biodiversity conservation and management organization, learn from the advanced experience of biodiversity conservation and management at home and abroad, learn from the specific measures of international famous scenic spots in the biodiversity management system, management mode, capital investment mechanism, etc., and continue to expand and deepen corrective actions in combination with the actual situation of scenic spots, so as to achieve a significant improvement in the level of biodiversity conservation and management in scenic spots.

Action 17: Strengthening the Protection of Ancient and Famous Trees

Carry out research on the management and rejuvenation technology of ancient and famous trees, and strictly implement regulations in the *Protection and Management of Ancient and Famous Trees in the Huangshan Scenic Area* and the *Regulations on the Rejuvenation Technology of Ancient and Famous Trees in the Huangshan Scenic Area*. Improve the protection plans for ancient and famous trees and the emergency protection plan for disastrous weather, strengthen the protection management measures, establish a comprehensive protection system, and strengthen the protection of ancient trees that are not within the scope of filing protection.

Action 18: Scientific and Reasonable Implementation of *Ex-Situ* Protection

Carry out breeding and restoration technology research of rare and endangered animal and plant species found at the scenic spot, focus on the artificial cultivation of special ornamental flowers, medicinal plants, and other species resources in Huangshan and the artificial cultivation of important protected animal species, so as to reduce the damage done to wild species resources. Establishing 1-2 *ex-situ* conservation parks, germplasm resource banks (nurseries), or important species rescue and breeding bases.

Action 19: Integrated Environmental Governance

①Carry out a comprehensive improvement of the environment in the scenic spot and improve the centralized disposal facilities for garbage and sewage.

②Integrate the treatment of sewage and garbage in villages and towns around the scenic spot, and gradually carry out the comprehensive treatment of sewage, garbage, agricultural non-point sources, and livestock breeding pollution in surrounding communities.

Priority Area VI: Accelerating the Scientific Utilization of Species Resources

Action 20: Strengthening Innovative Research on Scientific Utilization of Biological Species Resources

Accelerate the innovation research on the development and utilization technology and management mode of biological resources such as local characteristic crops, medicinal plants, ornamental flowers, aquatic organisms, and livestock genetic resources, so as to provide scientific basis for the sustainable utilization of biological resources in the scenic spot.

Priority Area VII: Sustainable Development of Local Communities

Action 21: Demonstration and Promotion of Alternative Livelihoods

Select rural communities around the scenic spot to establish demonstration pilot programs to explore ways of financial support and resource development. Select Huangshan tea, wild vegetables, medicinal materials, ornamental flowers, and other characteristic resources for market-oriented development according to local conditions. Set up a community characteristic commodity sales window in large-scale Huangshan themed activities. Vigorously develop eco-tourism and e-commerce, extend the tourism industry chain, promote the employment and entrepreneurship of community residents, promote the demonstration experience, and improve the overall development capacity of local communities.

Action 22: Demonstration and Promotion of Alternative Energy

Carry out the investigation of alternative energy and the study of technical and economic feasibility, select the surrounding communities as the pilot to carry out targeted demonstration and promotion of alternative energy, so as to reduce the pressure on vegetation, interference on ecosystem, and environmental pollution.

Priority Area VIII: Improving Policies, Regulations and Institutional Mechanisms

Action 23: Establishing and Improving Policies and Regulations Related to Biodiversity Conservation

①Improve laws and regulations related to biodiversity conservation, formulate and implement policies and measures that are conducive to biodiversity conservation, especially policies and measures that encourage investment and industry.

②Formulate and improve the ecological compensation mechanism for promoting biodiversity protection, study and formulate the ecological compensation mechanism and standards between resource development and ecological protection, between scenic spots and surrounding communities, and gradually establish a ecological compensation standard system for biodiversity protection.

③Formulate and promulgate the system regulations of biodiversity impact assessment that must be implemented in the EIA of major projects, and establish the public participation system of biodiversity impact assessment in the EIA of major projects.

④Formulate and issue the audit system for the departure of leading cadres' biological resources assets, the accountability system for biodiversity damage, and the damage compensation system.

Action 24: Innovating the System and Mechanism of Biodiversity Protection

①Study and establish the coordination mechanism of biodiversity protection, strengthen the cooperation among departments, promote information exchange and action coordination among all departments of the scenic spot, ensure smooth feedback channels of the relevant systems and policies of biodiversity protection, and establish and improve the cross departmental cooperation mechanism to combat the illegal acts that destroy biodiversity.

②Explore and establish the cooperation and co management mechanism between scenic spots and local communities.

Action 25: Strengthening Biodiversity Conservation in Sector Planning

①Promote the relevant departments of the scenic spot to incorporate the content of biodiversity protection into the development plan and work plan of the Department, and reflect the requirements of biodiversity protection.

②Establish the evaluation and supervision mechanism of the implementation of biodiversity conservation related plans, and promote the effective implementation of relevant plans of all departments.

Priority Area IX: Actively Respond to the Impacts of Climate Change on Biodiversity

Action 26: Climate Change Impact Assessment and Response to Biodiversity Conservation

①Study the impact of climate change on the important ecosystems, species, genetic resources, and related traditional knowledge of scenic spots, with an emphasis on the impact of climate change on biodiversity in sensitive areas.

②Explore the measures to deal with climate change, and formulate and constantly improve the biodiversity protection plan to deal with climate change.

③Local plant varieties with strong resistance shall be selected for greening in the scenic spot to enhance the ability of plants to adapt to climate change. Gradually establish a range of ecosystem sequence with reasonable structure and perfect function to mitigate the adverse effects of climate change on biodiversity.

Priority Area X: Promoting Publicity, Education and Public Participation

Action 27: Strengthening the Cultivation of Talent in the Field of Biodiversity Protection

①Develop and share biodiversity knowledge and professional technical training for management and technical personnel in relevant departments of the scenic spot, and improve their management level and professional technical ability.

②Establish a scientific and effective talent management mechanism to attract outstanding scientific research talent to join the scenic area for biodiversity conservation research.

③On the basis of rich biodiversity background data and scientific biodiversity protection level, actively attract talent and funds to Huangshan, and form a positive interaction of "talent funds technology".

Action 28: Enhancing Biodiversity Conservation Awareness and Education

①Carry out biodiversity publicity and education for tourists and community people through various media outlets such as Internet, newspaper, TV, radio, etc. Vigorously publicize the importance of biodiversity protection, improve the awareness of scenic area staff, tourists, and community residents on biodiversity protection. Advocate consumption patterns and catering culture conducive to biodiversity protection.

②Strengthen the legal popularization of biodiversity protection, and improve the legal awareness of biodiversity protection for managers, tourists, and local community residents.

③Support relevant local colleges, environmental protection associations, and social organizations to carry out publicity and education on biodiversity protection, organize the construction of ecological camps to carry out popularization and education of ecological knowledge, ecological experience, etc..

④Hold excellent biodiversity art works exhibition, encourage the creation of original biodiversity protection cultural and creative products in areas, such as film and television, photography, calligraphy, and painting.

Action 29: Establishing a Mechanism for Public Participation in Biodiversity Conservation

①Study the needs of the public in various aspects of biodiversity conservation, and establish channels and mechanisms for public participation in biodiversity conservation.

②Establish a multi-level partnership for biodiversity conservation, and guiding government departments, enterprises, the private sector, the public, and non-governmental organizations to participate in biodiversity conservation in various ways.

③Improve the level of intelligent technology for public participation, build an interactive platform for an animal and plant identification cloud in Huangshan, and use big data and artificial intelligence technology to improve public interest and awareness of biodiversity protection.

Action 30: Strengthening Foreign Exchange and Cooperation

Strive for international cooperation projects, introduce international advanced management experience, technology, and methods such as integrated ecosystem management, ecological compensation mechanism, biodiversity protection corridor construction, climate change, and biodiversity protection. Focus on international cooperation intelligent support projects. Actively establish a memorandum of friendship and cooperation with foreign protected areas to promote talent exchange and marketing.

6.3 Safeguard Measures

6.3.1 Organizational Leadership

Organize a lead biodiversity group for the Huangshan Scenic Area. The leading group of biodiversity is led by the director of the Management Committee of Huangshan Scenic Area, with the landscape Bureau of the Management Committee of Huangshan Scenic Area as the core department. Plan biodiversity management, coordinate the responsibilities of each department, and organize the implementation of the biodiversity emergency tasks.

Have the lead biodiversity group actively communicate and cooperate with other departments. Set the publicity department responsibilities for effective publicity of the biodiversity protection concept. The Discipline Inspection Commission's supervision office is responsible for the supervision of biodiversity conservation and respond to complaints from citizens and tourists. Give feed back to the responsible department. The land planning division is responsible for biodiversity conservation zoning and coordinating with the

Bureau of Landscape Architecture in the implementation of major projects. The economic development bureau is responsible for the development of the tourism area economy under the precondition of sustainable utilization of biological resources. The comprehensive law enforcement bureau is responsible for the compulsory implementation of the plan.

6.3.2 Supervision and Examination

Incorporate biodiversity conservation objectives into the work evaluation system. Through performance evaluation, the leadership is restricted to review and approve projects that are not conducive to biodiversity protection, and the leadership is urged to pay attention to biodiversity protection. Formulate the biodiversity protection responsibility scale, quantify the biodiversity protection achievements, and submit it to the Huangshan Municipal People's Congress for voting. The assessment results shall be provided with supporting reward and punishment measures. Work progress shall be reported regularly, relevant information shall be fed back, responsibility investigation for biodiversity protection shall be strengthened, and responsibility for illegal actions shall be investigated for those causing serious consequences. Illegal actions shall not be tolerated, and restriction barriers for biodiversity protection shall be formed for major decisions and major construction projects.

Implementing the strictest supervision system. The inspected unit or individual is required to provide relevant documents and materials to explain the work carried out, and may consult or copy them. The lead biodiversity group shall enter the crime scene for taking photos, video, and inspection. The group shall instruct member units or individuals to stop violating laws and regulations related to biodiversity protection. The group shall ensure the fulfillment of legal obligations. Monitor the areas where rare wild animals frequently appear and the areas where rare wild plants grow. Encourage the masses to participate in the work of biodiversity protection, resolutely investigate the illegal phenomena found and reported, and respond to the informants. Each case shall be reported and made public.

6.3.3 Fund Safeguard

Establishing the biodiversity conservation fund. Make use of the financial allocation to form a stable fund reserve to ensure the smooth development of various work related to biodiversity protection. Increase the government's investment in biodiversity protection to develop biodiversity protection informatization, scenic area management, endangered wildlife protection, natural vegetation protection and restoration, and invasive alien species prevention. Strengthen the management of funds for biodiversity conservation and improve the efficiency of their use. The project funds related to biodiversity conservation shall be audited and supervised to ensure the smooth progress of the project as planned.

Establishing multi financing channels. Make biological resource products market-oriented on the premise of sustainable utilization of biological resources in Huangshan according to the demand of the market. Adopt the PPP mode, e-commerce mode, P2P mode, etc. to introduce social funds, and more flexible policies in order to take full advantage of the market mechanism role in the allocation of ecological resources. Regularly release the financing intention of construction projects that may affect biodiversity protection. Guide social funds to flow to biodiversity protection. Take full advantage of the leading role of government investment and marketization. Make full use of multi-channel commercial financing means, raise social funds and expand the sources of funds.

6.3.4 Science and Technology Support

Strengthening the construction of talent echelon. Organize and carry out training related to biodiversity conservation, such as inviting experts to give lectures and allowing international advanced biodiversity conservation units to visit. Set up a biodiversity conservation think tank to provide accurate and professional technical support for the biodiversity conservation in Huangshan. Introduce biodiversity protection professionals through various channels, and create a good environment for the development of biodiversity protection talent by improving welfare treatment and creating professional promotion opportunities for them.

Setting up key scientific and technological projects. In view of the existing problems in Huangshan at this stage, carry out research, increase the investment of scientific research manpower and financial resources, and actively explore the effective ways and methods of biodiversity protection. Strengthen hardware investment, for biodiversity protection needs basic data investigation of various research subjects, which requires the establishment of biodiversity monitoring network mechanism to provide data support for in-depth research. Study climate change countermeasures. Understand the mechanism of habitat and climate change and the migration behavior of animals in the context of climate change in order to provide theoretical support for the protection of rare and endemic species in Mount Huangshan. Carry out research on the mechanism of forest pest control, to strictly control the invasion of alien species. Implement the early warning and control of pests.